朱雀文化

專業烤箱做料理 廚房從此零油煙

我曾在一間屋齡20多年，準備出售的老房子中，看見一台與廚房設備結合的崁入式烤箱，當時的我，看得好生羨慕。好奇打開烤箱，發現這台烤箱竟然在這個偌大的廚房裡，獨自孤單寂寞地度過它的歲月：烤箱內所有配件的塑膠封套還罩著，泛黃的使用手冊靜靜地躺在烤盤上。

多可惜啊！對我這個從小就愛上烤箱的人而言，烤箱給我的最大魅力，就是看著麵包或蛋糕經過加熱後慢慢膨脹的快感，快出爐時瀰漫在空氣中奶油和麥子的香甜氣味，是其他的廚房設備無法取代的。

然而事實上，烤箱不只是烤雞、烤魚或烤香腸，以同樣加熱的原理，烤箱也可以燉湯、煮飯、做料理。這一本食譜中特別規劃了幾道為人熟知的飯菜，例如港式臘腸煲飯、上海菜飯和油飯，也介紹了著名的西班牙海鮮飯。

餐後來一道甜點也是不可或缺的。因此書中介紹了10道點心，有中式的蛋黃酥，也有西式的布丁和蘋果派，數量雖然不及整本都是點心的食譜書，每一道卻也都是經過編輯和我再三討論後，覺得是經典之作才介紹給讀者的。

另外有趣的是，我們還教你用烤箱來燉湯。當您在家利用烤箱烹調晚餐時，可以先將需要長時間加熱的湯放入烤箱，約莫剩下1小時的時候，再放入米食類的食物與湯共同烹調，這樣不只可以省下不少時間和金錢，也不會產生有害人體健康的油煙。

記得有一次上廣播節目，與主持人夏韻芬小姐聊天，她是個善用烤箱的聰明主婦，懂得在宴客時，利用烤箱來烹調菜餚，再搭配一道有各式新鮮蔬菜的沙拉。這樣從容完成的宴客料理，讓女主人一手執鍋一手執鏟，攪動炒鍋內青菜，對著外頭客人拉開嗓子大喊：「你們先吃，我炒個菜就來！」的古早年代再也不會出現。

的確，不同的年代有不同的烹調手法，講究健康與無油煙的年代，就是需要一台專業的烤箱，來幫忙主婦料理美味餐點。

這次非常感謝由世磊實業公司提出拍攝一本烤箱菜的想法，並在食譜拍攝過程中全力配合，提供高品質高效能的烤箱，讓我過足了烤箱癮，真的很感謝世磊公司所有同仁的支持與鼓勵。再來是好友Lydia的全力幫忙，讓我可以從容地在預定的時間內完成50道食譜的製作工作，最高興的是再次與食譜攝影界的大師──徐博宇先生合作，讓我備感榮幸，再一次謝謝你們。

Foreword from the Author

I once saw, in a twenty year old house, a built-in oven in the kitchen. I was just heading out the door. At hat time, I really admired it. Out of curiosity I opened the oven and found the oven sitting there all alone, with all the accessories in a seal plastic bag and a yellowed instruction book still sitting quietly inside on the baking sheet. What a shame!

From my point of view, loving ovens since I was little, this was a great loss. Ovens have a great hold over me. I love to watch as a cake or bread swells in the oven, the air filled with the sweet-rich smell of wheat. Because of this experience, no other kitchen equipment compares to the oven.

In fact, ovens can not only be used to bake chicken or sausages, but can also be used to prepare soups, rice, and many other dishes. This cookbook has specially planned types of food that people already know, Hong Kong Style Sausage with Steamed Rice, Shanghai Style rice with Bok Choy and Oiled Rice, and I also introduce herein the famous Spanish Seafood Rice.

After meals, dessert is a must. This book introduces ten desserts as well, including Chinese-style Bean Paste with Egg Yolk Pastry, along with western Pudding and Apple pie. The number of dishes is not great, compared to whole cookbooks full of desserts, but each dish was picked after discussion with the editor, because it is a classic dish. We thus included it to introduce it to the readers.

Another interesting aspect of the cookbook is that we show you how to make soup with the oven. When you make dinner, place soups that require long cooking times in the oven. When an hour of cooking time remains, add other dishes. This saves time and money, and does not produce the smoke associated with frying, which is unhealthy.

I remember once when I was on a TV show chatting with its host, Miss Hsia Yun-fen, a very smart housewife who knows how to make good use of the oven. Whenever people come over her house, she uses the oven to prepare the dishes, accompanied by a fresh salad. This way, you can complete your meal preparations in a relaxed manner. In the old days, the wife always cooked with one hand stirring the wok, yelling to the guests sitting outside: “Hey! You guys start eating! Just let me finish this last dish.” This will never again.

It is true that each era has its own cooking methods. In this healthy, no-smoking era, you need an oven to help the homemaker cook delicious meals.

This time around I want to express my deep appreciation to Shih Ye Company, which suggested oven cooking as a topic, accompanied by plenty of pictures. During the three days of shooting for the cookbook, they put all their energy into getting it done right. They offered a high quality, high efficiency oven, fully satisfying my own addiction to ovens. I truly appreciate the support and encouragement of their enthusiastic staff. I would also like to thank my good friend Lydia for her dedication and help, which enabled me to complete the fifty dishes within the designated days. I am happiest to note that once again, my cookbook photographer was Mr. Hsu Po-yu, a master of cooking photography. I am deeply honored to have the opportunity to thank you all once again.

Contents

烤箱新手的第一本書

Part2 主食和菜餚 *Main Courses & Dishes*

Part3 中西式點心 *Chinese dessert &Western dessert*

馬上學會用烤箱
Learn how to cook with oven

使用本書之前的小指南

一個專業的烤箱，基本上都會有傳統調理及紅外線烹煮等三種高溫火力的多重控溫選擇，想要製作美味的烤箱菜，一台專業的烤箱是不可或缺的。

每道食譜我們都附有專業烤箱使用的簡易功能小表格，方便你器具的準備與烤箱溫度的控制。

Special Tips before Reading

A professional oven basically contains traditional controls, along with a UV sensor, to control three types of temperature selection. In order to prepare delicious dishes, a professional oven is a must.

We attach a simple form for professional use of the oven to every recipe, so that the readers can prepare the utensils beforehand and easily control the temperature of the oven.

溫度與器具標示僅供參考，請依據實際烤箱為準，建議你多試試自己的烤箱、熟悉烤箱功能，才能更愉快的做菜喔！

The temperature and utensil labels are only for reference. Each make of oven has different temperature settings. I suggest you experiment with the oven you own, and become familiar with its behavior and functions in order to prepare a successful meal.

選購烤箱之前

1.確認預算與需求買烤箱

購買烤箱的第一步，要先確認自己的預算，在預算範圍內去選擇最適用的品牌。有些品牌著重在經典的外觀，容量空間小，售價卻不便宜；有些品牌的外觀雖然很普通，但是功能卻很實用，例如同時結合烤箱和微波的雙重機型烤箱、可以在烘烤時噴灑蒸氣的蒸氣式烤箱。而進口烤箱的操作面板全都以英文或圖案顯示，對於不懂英文的人也許不夠方便，但進口烤箱功能齊全，甚至還可設定安全上鎖，這點非常安全便利。

2.完善的售後服務

購買烤箱時要格外確定有無售後服務以及機器的保固期。因為進口烤箱多來自較寒冷的歐美國家，台灣因為地處潮濕的亞熱帶，氣候不同，有可能造成機器零件的壽命縮短或故障，因此完善的售後服務非常重要。

3.烤箱擺設位置要完善

烤箱在加熱時會產生高溫，所以烤箱的位置最好與料理平台高度相等，可以避免家中年幼的小孩不甚碰觸而燙傷，也可免去操作者要不斷彎腰與蹲下查看。

如果是為了搬遷新居而選購烤箱，此時應與設計師討論，或是請烤箱廠商提供內部設計師，為廚房整體空間做最完善的規劃。規劃的重點除了擺設位置外，還包括電壓與散熱等問題。

4.選擇功能齊全的烤箱

每次進行現場點心教學時，都會遇到的一個問題是：烤吐司和麵包用的小烤箱可不可以用來烤蛋糕？答案是：「不可以。」因為小型烤箱的溫度太高，空間太小，很容易將蛋糕表面烤焦，但是蛋糕內部卻尚未熟透。

因此，選購一台適合自己和家人使用的烤箱，是最重要的第一步，再來就是放心大膽地使用它，依照食譜書上記載的時間和溫度，製作你喜歡的料理，只要多試幾次，就可以得心應手地掌握了。

一個功能齊全的烤箱應具備以下的功能：

烤箱模式 Oven Mode

上下火、受熱均勻的烘烤模式。
Upper and lower heating elements with balanced cooking modes.

解凍模式 unfreeze the way

啓動烤箱風扇功能，以常溫解凍食材。
Start the fan function of the oven, unfreeze and eat the material with the normal atmospheric temperature.

排餐模式 Steak Mode

啓動大面積點火的加熱，均勻的使食物與爐面接觸，適合排餐及吐司。
Turns on the large area heat, so that the food and the baking sheet are evenly heated. Suitable for steaks and toast.

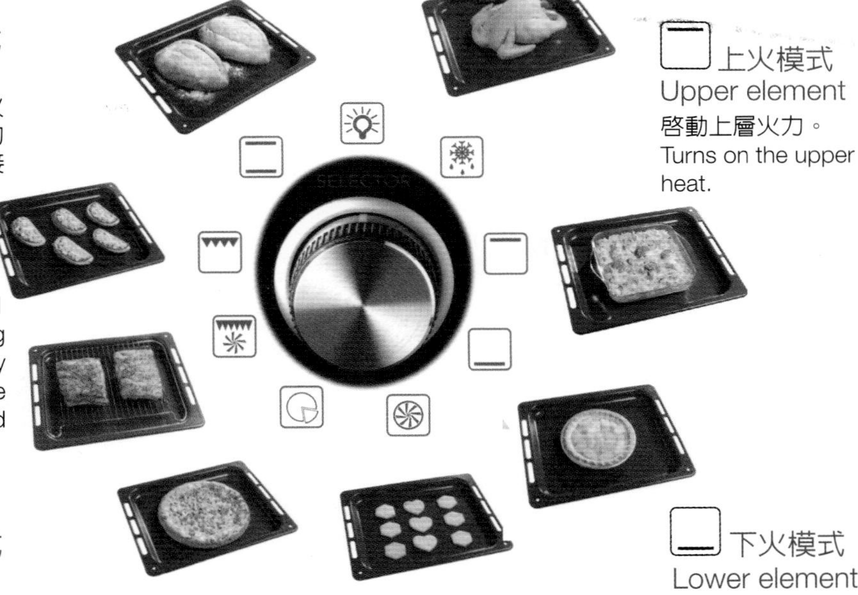

上火模式 Upper element

啓動上層火力。
Turns on the upper heat.

燒烤模式 Baking and Roasting Mode

啓動大面積點火的烤箱加熱，適合排餐及厚實的肉塊。
Turns on large area heat, suitable for steaks or other thick meat dishes.

披薩模式 Pizza Mode

啓動環型加熱、下層點火及烤箱風扇的功能。
Turns on even, all-around heat, lower heat and fan.

風扇模式 Fan Mode

加熱烤箱內的空氣，適合多層口味的菜餚。
Heats up the air in the oven, suitable for dishes with layers.

下火模式 Lower element

啓動下層火力。
Turns on the lower heat

正確使用烤箱

1.熟讀操作手冊

不論你購買的烤箱是哪種。根據我多年的經驗，熟讀操作手冊是重要的第一步。因此一定要善加運用手冊，並且好好保存它。

2.溫度的設定

一般的烤箱都會標示℃（Celsius）或℉（Fabrenbeit），也就是攝氏或華氏，很多家庭主婦至今仍然搞不清楚怎麼區分，一般歐美進口的烤箱都以華氏（℉）標示烤箱溫度，而日製及台製烤箱皆以攝氏（℃）標示，所以在你買回烤箱後，要確認你所擁有的烤箱是使用何種度數標示。以下是攝氏及華氏換算表：

攝氏（℃）/華氏（℉）

100/210，120/250，150/300，175/350，190/375，200/395，220/425，230/450

3.正確的設定時間

由於每部烤箱的不同，時間上也會有所差異，同時食物的厚薄、使用模子的大小，在時間上也必須略做調整，一般而言烘焙較厚的食物時，時間要拉長、溫降要度低，而烘焙較薄的食物時，時間要減少、溫度要升高。若是你的烤箱沒有時間鍵的話，最好能自己準備一個定時器，以免因一時的疏忽而將食物烤焦。假如你對於你的烤箱性能還不是很了解的話，最好先設定較短時間，覺得不夠熟，再延長時間，若所烤的時間超過本食譜所設時間10分鐘以上，則你就要加高烘焙的溫度。

烤箱的保養與清潔

在正式啓用全新的烤箱之前，可先以乾淨的濕布或廚房紙巾將烤箱內外仔細擦拭一遍，接著按照操作手冊進行第一次加熱。通常手冊上會建議消費者不放食材，讓烤箱空烤10至15分鐘，此時若有微量的白色煙霧冒出，應屬正常；如果有異味自烤箱內竄出，請開啓烤箱門散熱即可，接下來就可按照正常的使用方式來啓用。

2.清潔前先拔插頭

清潔烤箱的時候請先拔除插頭。如果烤箱內壁已沾染油漬，可先用軟性清潔劑噴灑在抹布上，再以抹布擦拭油漬沾染處，切勿直接將清潔劑的噴頭對著油漬噴灑，以免破壞烤箱的電熱管線。

3.烤箱各部位清潔法

烤盤：每次使用完畢，都應該將烤盤取出，放置在洗碗槽或洗碗機內清洗，清洗後立刻以乾抹布或廚房紙巾擦乾，盡量避免濕濕的晾乾，可以確保烤盤不生鏽。

內外、手把和按鍵或是旋轉鈕：若每一次使用結束後都加以擦拭，可以延長烤箱外觀的壽命。

清潔電熱管線：清潔時，務必拔除插頭，並等待其完完全全降溫，再以乾淨的抹布或軟鬃刷來清潔，切勿使用清潔劑或水。

4.定期保養烤箱

烤箱和人一樣都需要保養，建議大約每隔1至3個月的時間，進行一次保養動作。方法很簡單，就是將有深度的烤盤注入水，放入烤箱開啓加熱，熱度大約是100℃，時間設定為15至20分鐘，讓烤箱內充滿蒸氣；隨後拔除插頭，並等待烤箱降溫，再以乾抹布擦拭即可。平常如果因為製作味道較重的料理而留下味道，也可以用上述方法來清除味道，若是在烤盤上放鳳梨皮、柑橘皮或檸檬皮，也可以幫助消除異味，效果很好。

Before selecting an oven

1. Estimate the budget and needs for a new oven

The first step in buying an oven is to determine your own budget and select the most suitable brand without the budget limit. Some brands emphasize classic looks with smaller interior space, yet the price is not cheap. Some brands, though common looking, have practical options, for example, dual-functionality which combines the functions of oven and microwave and the steaming style oven, which sprays steam during baking. Imported ovens with controls written in English or shown in photos, for those who do not understand English, are not very convenient. However, imported ovens have a wide range of functions, some even have safety locks, safe and convenient.

2.Complete after sales service

When buying a new oven, be sure that after sale services is offered, along with a warranty for the machine, because most of the imported ovens come from European countries with colder climates. Taiwan is located in the humid subtropical zone, and different climates may possibly shorten the life of machine parts or cause breakdowns. Therefore complete after sale service is very important.

3.The location of the oven

When an oven is used, it produces high temperatures, so the location of the oven should be at the same height as the cooking counter, to prevent toddlers from getting burnt due to carelessness or curiousity. This also prevents the operator from bending and kneeling to check the oven and its contents.

If you purchase ovens due to moving, a discussion between you and the designer should be conducted, or have the oven factory offer you an interior designer for the kitchen space. The essence of the designing, in addition to location of the oven, also includes such items as voltage, heat removal, and so on.

4.Select ovens with a wide range of functions

Every time we conduct a dessert class, we encounter the question: Can the toaster oven be used for baking cakes? The answer is no, because the temperature of a toaster oven is too high and the space is too limited. It is too easy to burn the cake on the surface with the inside still uncooked.

Thus, selecting an oven that is suitable for yourself and your family is the most important priority. The next step is to get the most out of it, following the times and temperatures written in the recipes, to prepare the dishes you like. If you keep trying, you can manage it in no time.

How to use an oven correctly

1.Read instruction menu thoroughly

No matter what brand of oven you buy, based on my many years experience, the most important first step is to read the operating instructions thoroughly. Make good use of the instructions, and maintain it well.

2.Setting the tempernture

Most ovens are labeled in either°C (Celsius) or°F (Fahrenheit). Most European and American ovens use Fahrenheit, while Japanese and Taiwanese label in Celsius. When you bring your oven home, check to see what it uses. Below is a conversion table for Celsius and Fahrenheit.

Celsius (°C) / Fahrenheit (°F)

100/210 120/250 150/300 175/350

190/375 200/395 220/425 230/450

3. Setting time correctly

Adjust the timer according to the thickness of the food, and the size of the pan. Generally speaking, thicker food demands more baking time and lower temperatures, thinner food needs less time and higher temperatures. If your oven doesn' t include a timer, be sure to get one. If you are not familiar with your oven, set a shorter time first, if the food is not done when the timer goes off, prolong the time. If the baking time is 10 minutes more than the book says, increase the temperature.

The maintenance and cleanliness of an oven

1.Bake empty the first time

Before starting to use your oven, clean with a damp cloth or paper towel in and out thoroughly, following the instructions, and carry out the initial heating. Normally the instructions suggest the consumer let the oven bake for 10 to 15 minutes without any food inside it. At this time a small amount of white smoke will emerge, a normal phenomenon. If there is any strange smell coming out, just open the oven door to release the heat. Now you can start using the oven in the normal way.

2.Unplug before cleaning

Remove the plug before cleaning the oven. If there is stain on the inside wall of the oven, spray detergent or cleanser on the kitchen cloth, then use the cloth to mop up the stains. Do not spray the detergent directly onto the stain, to prevent it from destroying the electric elements inside the oven.

3.How to clean parts of the oven

Baking Sheet : Baking sheet should be removed after each use. Clean it in the sink or dishwasher, then remove and dry well with a dry kitchen cloth or paper towel. Avoid air drying to prevent rust.

Inside and Outside, Handle & Buttons or Knobs : Swab thoroughly each time after baking to prolong the life of the oven and its appearance.

Electric Elements : Be sure to unplug when cleaning and wait until the oven completely cools down, then clean with a dry kitchen cloth or soft oven mane brush. Do not use detergent or water.

4.Maintenance steadily

An oven is just like a human, and needs maintenance. Every one to three months the following maintenance is suggested. The methods are simple: just fill the baking sheet with water and place in the oven. Turn the heat to approximately 100°C and set the time to 15 to 20 minutes to let the oven fill with steam. Next unplug and wait until the temperature drops, then mop with a dry kitchen cloth. When we prepare dishes that leave the oven with a strong odor, the methods above can be used to clear up the odor. Or placing pineapple skin, orange peel or lemon peel on the baking sheet is also effective in clearing up the odor.

● 感謝世磊實業提供義大利 best 烤箱資訊及圖片
(02) 2762-6096

Side Dishes & Soup

配菜和湯

從基礎做法到進階料理，從創意配菜到香濃湯品，烤箱教你神奇做好菜！

西班牙早餐蛋

Spanish Breakfast Eggs

烤箱用 烤/ 模具 cookware	烘烤功能 Functions	溫度℃	時間 min
耐熱烤皿x2 Heat Resistant Baking Utensils x2	or	150	25

材料

雞蛋2顆、早餐腸150公克、塗抹用奶油1大匙

❶ 去皮去籽蕃茄丁1杯、洋蔥碎2大匙、迷迭香1小匙、橄欖油2大匙

❷ 蕃茄醬3大匙、白酒3大匙

❸ 鹽1小匙、黑胡椒粉1/2小匙

做法

1. 早餐腸斜切3公分小段。
2. 將橄欖油倒入平底鍋中，以中火將材料❶炒軟。
3. 然後加入材料❷煮至略收汁，起鍋前加入材料❸調味。
4. 此時開啟烤箱預熱。
5. 準備兩個烤皿，內部塗抹奶油。將炒好的料平均倒入烤皿，用湯匙各挖出一個深可見烤皿底部的洞，各打入一個雞蛋。
6. 最後將早餐腸切小段，排在雞蛋周圍，稍微將材料壓入，即可放入烤箱，以上下火150℃烘烤約25分鐘即可。

Tips　材料中使用的去皮去籽蕃茄丁，是方便好用的進口罐頭食品。如果不喜歡罐頭食物，也可以直接將新鮮牛蕃茄汆燙後剝皮去籽，再將果肉切小丁即可。

Ingredients

2 eggs,150g breakfast sausages 1T. butter, parsley for garnishing as desired

❶ 1 cup peeled,seedless diced tomato, 2T.chopped onion, 1t. rosemary, 2T.olive oil

❷ 3T.ketchup, 3T.white wine

❸ 1t. salt, 1/2t. ground black pepper

Methods

1. Cut breakfast sausages into sections 3cm. long.
2. Pour olive oil in a frying pan and stir-fry ingredient ❶ over medium heat until softened.
3. Add ingredient ❷ and cook until the liquid is almost absorbed. Season with ingredient ❸ before removing from heat.
4. Preheat oven.
5. Prepare two baking bowls and grease the inside with the butter. Spread method evenly around the bowl. Dig a hole in the center of each bowl deep enough to see the bottom, and break an egg inside each hole.
6. Surround the sides of the eggs with the breakfast sausage sections and press the sausages in slightly. Remove to the oven and bake at with both elements at 150℃ for 25 minutes until done.

Tips

Using imported canned peeled seedless diced tomato is very convenient and saves lots of work. If not desired, just blanch a fresh beefsteak tomato first and peel the skin, then remove seeds and dice.

Spanish Breakfast Eggs

青辣紫茄起司盤

Spicy Eggplant in Cheese Dressing

烤箱用 烤/ 模具 cookware	烘烤功能 Functions	溫度℃	時間 min
耐熱烤盤 Heat Resistant Baking Sheet	or	175	15

材料

茄子3支、橄欖油3大匙、鹽2小匙、軟質的瑪芝瑞拉起司200公克、裝飾用九層塔少許

❶青辣醬料：青辣椒75公克、紅辣椒25公克、九層塔2束、橄欖油2大匙

做法

1. 將茄子切成四等分，每一等分再縱向剖成兩半。白色茄肉的部份以刀子輕輕劃斜線，幫助醬料滲入，此時開啓烤箱預熱。
2. 將茄子放在耐熱烤皿內，表面淋上橄欖油並撒上鹽，放入烤箱以175℃烘烤15分鐘。
3. 青辣椒和紅辣椒去蒂頭後切碎，九層塔取葉子的部份切碎，把兩種材料混合再切碎，加入橄欖油拌勻，即成青辣醬料（可改成以小型食物研磨機代替）。
4. 起司切成0.5公分薄片，取出烤好的茄子，與起司交叉排列，淋上青辣醬料，並以九層塔葉點綴即可。

Tips 可選擇日本茄子，其外型圓胖，切片後呈圓片狀，排盤更漂亮且口感更實在。

Ingredients

3 eggplants, 3T. olive oil, 2t. salt, 200g soft mozzarella cheese, basil for garnishing

❶ 75g green chili pepper, 25g red chili peppers, 2 bunches basil, 2T. olive oil

Methods

1. Cut each eggplant into 4 equal sections, then cut each section in two lengthwise. Score diagonally on the inner white flesh a few times to help the seasonings be absorbed easily. Preheat oven.
2. Remove eggplants to a baking bowl and drizzle with olive oil, then sprinkle with some salt and place in oven. Bake at 175℃ for 15 minutes until done.
3. To prepare the chili dressing: Remove stems from green and red chili peppers, then mince well. Retain basil leaves and chop finely. Combine these above ingredients together and chop finely once more. Add olive oil to mix. (The procedure can be accomplished with a small food processor.)
4. Cut cheese into slices 0.5cm thick. Remove eggplants and arrange on a serving plate alternating with the cheese slices, then drizzle with the chili pepper dressing and garnish with basil leaves. Serve.

Tips

Fat round Japanese eggplants can be used in this dish, because the slices are round and look pleasing when arranged on a serving plate. The texture is delicious.

Spicy Eggplant In Cheese Dressing

烤箱用 烤/ 模具 cookware	烘烤功能 Functions	溫度℃	時間 min
耐熱烤盤 Heat Resistant Baking Sheet	 or 	100、180	5～7、25

烤地瓜佐鮮蔬沙拉

Baked Yam with Salad

材料

地瓜1個、綜合綠葉生菜100公克、酸奶油2大匙、原味優格100公克、核桃粒2大匙

❶ 鹽1小匙、黑胡椒粉1/2小匙、檸檬汁1小匙、橄欖油2大匙

做法

1. 將核桃粒放在小烤盤上，放入烤箱以100℃烘烤5至7分鐘。
2. 地瓜表皮洗淨擦乾，整顆放入烤箱，以180℃烘烤25分鐘。
3. 將酸奶油和優格混合拌勻，再加入材料❶即成酸奶優格醬。將烤好的地瓜取出，去皮切塊擺入盤中，淋上酸奶優格醬，並搭配洗淨的生菜和烤過的核桃即可趁熱食用。

Tips 這道料理的另外一個做法，是將烤熟的地瓜與核桃混合拌成泥，再搭配少許葡萄乾，口感也不錯。

Ingredients

1 yellow yam,100g mixed green salad greens,2T. sour cream,100g plain yogurt, 2T. walnuts

❶ 1t. salt,1/2t. ground black pepper,1t. lemon extract,2T. olive oil

Methods

1. Roast walnuts on a small baking sheet in oven at 100℃ for 5 to 7 minutes.
2. Rinse yam well and bake in oven at 180℃ for 25 minutes.
3. Combine sour cream and yogurt until even, then add ingredient ❶ to make sour cream yogurt sauce. Remove the yam from oven and split with a knife and drizzle sour cream yogurt sauce over the dish. Serve with rinsed greens and roasted walnuts before yam gets cold.

Tips

Another way of preparing this dish, which is also very delicious, is to combine the cooked yam with walnuts, then serve with raisins as desired.

烤箱用 烤/ 模具 cookware	烘烤功能 Functions	溫度°C	時間 min
耐熱烤皿 Heat Resistant Baking Utensils		175	15

烤綜合鮮香菇

Baked Mushroom Combination

材料

杏鮑菇150公克、鮮香菇150公克、蘑菇100公克

❶ 橄欖油2大匙、鹽1大匙、黑胡椒粉1小匙

❷ 黑醋2小匙、芫荽葉2大匙

做法

1. 所有材料洗淨瀝乾，放入烤皿，此時開啓烤箱預熱。
2. 加入材料❶拌勻，放入烤箱以175℃烘烤15分鐘。
3. 取出烤好的材料，撒上材料❷即可趁熱食用。

Tips 香菇盡量選擇肉質厚實的品種，烘烤時才容易翻動，讓受熱均勻；芫荽就是台式香菜，亦可以省略不用，或是使用九層塔或細香蔥來取代。

Ingredients

150g abalone mushrooms, 150g fresh shiitake mushrooms, 100g cremini mushrooms

❶ 2T. olive oil, 1T. salt, 1t. ground black pepper

❷ 2T. black vinegar, 2T. cilantro leaves

Methods

1. Rinse all the mushrooms well and drain thoroughly, then remove to a baking sheet. Preheat the oven.
2. Sprinkle with ingredient ❶ and mix well. Bake in oven at 175℃ for 15 minutes until done.
3. Remove and sprinkle with ingredient ❷ to mix. Serve while still steaming.

Tips

Select thick, firm shiitake mushrooms for this dish, so that it is easier to turn them over and evenly heat them when baking. Cilantro can be omitted, or substituted with basil or shredded scallions.

法式酥皮火腿+水蜜桃起司烤厚片吐司

Baked French Style Pastry over Ham Toast & Peach Cheese Toast

烤箱用 烤/ 模具 cookware	烘烤功能 Functions	溫度°C	時間 min
耐熱烤盤 Heat Resistant Baking Sheet	or	220	5～7

材料

厚片吐司2片、鹹味奶油2大匙、蛋液適量

❶ 法式酥皮火腿：

冷凍酥皮1片、綠葉生菜1片、早餐起司1片、蕃茄2片、法式火腿2片

❷ 水蜜桃起司：

罐頭水蜜桃1個、早餐起司1片、綠葉生菜1片、蕃茄1片

做法

1. 在厚片吐司表面塗抹鹹味奶油，於此時開啟烤箱預熱。
2. 法式酥皮火腿：依序在吐司上放入生菜、起司、蕃茄和火腿，最後蓋上冷凍酥皮，並在表面均勻塗抹蛋液。
3. 水蜜桃起司：水蜜桃切薄片（不切斷），切成類似香蕉的外型，用廚房紙巾吸乾多餘湯汁。厚片吐司表面依序放入生菜、起司和水蜜桃。
4. 將兩片吐司放在烤盤上，放入烤箱，以上下火220℃烘烤5至7分鐘。

Tips 也可改用披薩起司來取代冷凍酥皮，此外也可以將材料任意配成玉米鮪魚、培根蕃茄、新鮮水蜜桃或蜜李等自己喜愛的口味。

Ingredients

2 thick slices bread, 2T. salted butter egg as needed

❶ 1 frozen puff pastry sheet, 1 green lettuce leaf, 1 slice breakfast cheese, 2 slices tomato, 2 slices French style ham

❷ 1 canned peach, 1 slice breakfast cheese, 1 green lettuce leaf, 1 slice tomato

Methods

1. Spread the salted butter evenly over the surface of the bread and preheat the oven.
2. To prepare French style pastry over ham: Place lettuce leaf, cheese slice, tomato and ham in that order on the bread, top with the frozen pastry sheet and spread the egg liquid evenly across the surface.
3. Peach Cheese Toast: Cut canned peaches into thin, banana-shaped slices. Be sure not to break them. Remove the liquid by daubing with a paper towel. Top the other slice of bread with lettuce leaf, cheese and peach.
4. Place the two slices of bread on a baking sheet and bake in oven with upper and lower element both at 220℃ for 5 to 7 minutes until done.

Tips

Frozen pastry sheets can be substituted with pizza cheese. Ingredients such as corn and tuna, bacon and tomato, fresh peach or plum, or any preferred flavor, may also be used.

Baked French Style Pastry over Ham Toast & Peach Cheese Toast

法式田園蔬菜塔

French Style Garden Vegetable Tart

烤箱用 烤/ 模具 cookware	烘烤功能 Functions	溫度°C	時間 min
耐熱烤盤 Heat Resistant Baking Sheet	or	180	25

材料

❶ 塔皮：無鹽奶油120公克、糖粉25公克、細砂糖25公克、全蛋1顆、低筋麵粉250 公克

❷ 蔬菜餡：洋蔥50公克、蘆筍50公克、玉米粒1大匙、巧達乳酪絲50公克、橄欖油1大匙、鹽1小匙

❸ 蛋塔水：牛奶125公克、鮮奶油125公克、全蛋2顆、蛋黃1顆、起司粉少許、鹽1/2小匙、黑胡椒粉1/4小匙

做法

1. 塔皮製作：將無鹽奶油、糖粉和細砂糖混合拌勻。
2. 全蛋加入拌勻。
3. 低筋麵粉加入拌勻。
4. 拌成麵糰後，以保鮮膜包裹，移入冰箱冷藏至少2小時。
5. 塔模塗油撒粉，塔皮麵糰取出退冰後擀平，放入模型中整平，並以切麵刮刀切除多餘塔皮。
6. 製作蔬菜餡：用橄欖油將洋蔥、蘆筍炒至半熟，加鹽調味關火，再拌入玉米粒和乳酪絲即成餡料。
7. 製作蛋塔水：將全蛋和蛋黃放入鋼盆拌勻，加入鮮奶油和牛奶拌勻，透過濾網過濾蛋液，再加入起司粉、鹽和黑胡椒粉拌勻，即成蛋塔水。
8. 將炒過的蔬菜放入塔皮，加蛋塔水至全滿，撒上1大匙乳酪絲在表面，放入烤箱下層，以180℃烘焙25分鐘。

Ingredients

❶ Tart: 120g unsalted butter, 25g confectioner's sugar, 25g fine granulated sugar, 1 whole egg, 250g cake flour

❷ Vegetable Stuffing: 50g onion, 50g asparagus, 1T. corn kernels, 50g shredded cheddar cheese, 1T. olive oil, 1t. salt

❸ Tart liquid: 125g milk, 125g whipping cream, 2 whole eggs, 1 egg yolk, cheese powder as needed, 1/2t salt, 1/4t black pepper powder

Methods

1. To prepare tart: Combine unsalted butter, confectioner's sugar and granulated sugar together.
2. Add whole eggs.
3. Fold in cake flour to mix.
4. Mix well to form a soft dough, wrap up in saran wrap and chill in the refrigerator for at least 2 hours.
5. Grease the tart mold and sprinkle with flour beforehand. Remove tart dough from the refrigerator and defrost, then roll flat with a rolling pin. Place the dough skin over the mold, press down in mold and even up the edges, then trim off any excess dough.
6. To prepare vegetable stuffing: Stir-fry onion and asparagus with olive oil until half done, add salt to taste and remove from heat. Stir in corn kernels and shredded cheese, then mix well.
7. To prepare tart liquid: Mix whole eggs and egg yolk in a mixing bowl well, add whipping cream and milk. Stir until even, then pour through a sifter to remove any dregs, then add cheese powder, salt and black pepper to taste.
8. Pour vegetable stuffing inside the tart, then fill with the tart liquid until full, sprinkle with 1 tablespoon of shredded cheese on top. Bake at the lower level of the oven at 180℃ for 25 minutes until done. Serve.

French Style Garden Vegetable Tart

羅勒醬烤蕃茄

Baked Tomato with Basil Sauce

烤箱用 烤/ 模具 cookware	烘烤功能 Functions	溫度°C	時間 min
耐熱烤皿 Heat Resistant Baking Utensils	or	175	15

材料

牛蕃茄5個、市售羅勒醬4小匙、橄欖油1大匙、法國麵包1條

做法

1. 牛蕃茄洗淨，每個切成四片，再將籽挖除，此時開啓烤箱預熱。
2. 法國麵包切片。
3. 準備烤皿，淋上橄欖油，將蕃茄擺入，於每顆蕃茄表面舀入適量的羅勒醬，移入烤箱以175℃烘烤15分鐘，法國麵包亦放入烘烤。
4. 取出烤好的麵包，表面塗抹份量外的橄欖油。烤好的蕃茄去皮、切條狀，放在麵包上面，即可趁熱食用。

Tips 另一個做法是將烤好的蕃茄去皮後切碎，拌入橄欖油、紅酒醋和蒜末，食用時舀一小匙塗抹在麵包上面即可。

Ingredients

5 beefsteak tomatoes, 4t. market-sold basil sauce, 1T. olive oil, 1 stick French bread

Methods

1. Rinse beefsteak tomatoes well and cut each horizontally into 4 equal slices, then discard seeds, Preheat oven.
2. Cut French bread into slices.
3. Prepare a baking sheet and drizzle with olive oil. Place tomatoes in sheet and top with suitable amount of basil sauce. Place in oven and bake at 175℃ for 15 minutes until done. Bake French bread in oven until done.
4. Remove bread and spread olive oil across the surface evenly. Peel off the skin from tomatoes and cut into strips. Add to the bread and serve while it is still steaming.

Tips

Another way of preparing this dish is to bake the tomatoes in oven first, remove skin and chop finely. Mix well with olive oil, red wine vinegar and minced garlic. Spread a small teaspoon over bread when serving.

Baked Tomato with Basil Sauce

培根烤馬鈴薯

Baked Potato with Bacon

烤箱用 烤/ 模具 cookware	烘烤功能 Functions	溫度°C	時間 min
耐熱烤皿 Heat Resistant Baking Utensils	or	200	20

材料

馬鈴薯3個、培根3條、披薩起司30公克、青蔥末3大匙、酸奶油60公克、鹽3小匙、黑胡椒粉3小匙

做法

1. 將馬鈴薯整顆洗淨，放入鍋中，倒入足夠蓋過馬鈴薯的水，以中火煮約10分鐘。
2. 取出馬鈴薯切半，放入烤皿。
3. 培根切碎平均撒在馬鈴薯表面，起司也撒在馬鈴薯表面，此時開啓烤箱預熱。
4. 將烤皿放入烤箱，以200℃烘烤20分鐘，取出烤好的馬鈴薯，表面撒上青蔥末、鹽和黑胡椒粉，並搭配酸奶油即可趁熱食用。

Tips 選擇嫩皮馬鈴薯可以省去削皮的動作，也可以保留更多養分！

Ingredients

3 potatoes, 3 strips bacon, 30g pizza cheese, 3T. minced scallion, 60g sour cream, 3t. salt, 3t. ground black pepper

Methods

1. Rinse the whole potatoes well and cook in pan with water enough to cover all the potatoes, then cook over medium heat for about 10 minutes.
2. Remove potatoes and cut them in halves, then remove to a baking sheet.
3. Chop bacon strips into pieces, and sprinkle evenly over potatoes, then sprinkle cheese evenly over top. Preheat the oven.
4. Place baking sheet in the oven and bake at 200℃ for 20 minutes. Remove the cooked potatoes and sprinkle with minced scallion, salt and ground black pepper. Serve right away with sour cream.

Tip

Select potatoes with tender skin, which saves peeling and retains nutrients.

Baked Potato with Bacon

烤箱用 烤/ 模具 cookware	烘烤功能 Functions	溫度℃	時間 min
耐熱烤皿 Heat Resistant Baking Utensils	or	200	25

奶油焗馬鈴薯

Baked Potato Slices with Butter

材料

奶油30公克、低筋麵粉30公克、牛奶300公克、馬鈴薯400公克、披薩起司75公克、帕米森起司粉2大匙

❶ 鹽2小匙、黑胡椒粉1小匙、肉豆蔻粉1/2小匙

做法

1. 將奶油放入鍋中，以小火加熱融化，倒入麵粉以網狀攪拌匙拌勻成麵糊，倒入牛奶繼續攪拌至濃稠且沸騰，加材料❶調味後關火，即成奶焗醬。
2. 把奶焗醬倒入烤皿，此時開啓烤箱預熱。
3. 馬鈴薯去皮切片，放入滾水汆燙3分鐘，撈起瀝乾排入烤皿，表面撒上披薩起司，放入烤箱以200℃烘烤25分鐘，取出烤好的馬鈴薯，表面撒上帕米森起司粉，即可趁熱食用。

Ingredients

30g butter, 30g cake flour, 300g milk, 400g potato, 75g pizza cheese, 2T. parmesan cheese powder,
2t. salt, 1t. ground black pepper, 1/2t. nutmeg

Methods

1. To prepare the butter cream sauce: Melt butter in pan over low heat, add some flour and stir with a wire mixing spoon evenly to form batter, then pour in milk and continue stirring until thick and boiling. Add ingredient❶ and remove from heat.
2. Preheat oven. Pour the butter sauce in the baking sheet.
3. Peel potato and cut into slices, rinse out the starch from surface and drain well. Arrange on baking sheet and sprinkle the surface with pizza cheese. Bake in oven at 200℃ for 25 minutes until done. Remove and sprinkle with parmesan cheese. Serve right away.

烤箱用 烤/ 模具 cookware	烘烤功能 Functions	溫度°C	時間 min
耐熱烤皿 Heat Resistant Baking Utensils	or	200	30

奶油烤玉米
Buttered Corn

材料

黃玉米4支、鹹味奶油4大匙、鹽1小匙

做法

1. 奶油軟化後拌入鹽，並準備毛刷，此時開啓烤箱預熱。
2. 用毛刷將軟化的奶油塗抹在玉米表面，再以鋁箔紙將玉米包緊，放在烤盤上，送入烤箱，以200℃烘烤30分鐘。取出烤好的玉米，剝除鋁箔紙即可趁熱食用。

Ingredients

4 ears of sweet corn, 4T. salted butter, 1t salt

Methods

1. Soften butter and add salt to taste. Prepare a brush and preheat oven.
2. Brush the butter over the corn and wrap up well with aluminum foil. Place on a baking sheet and bake in oven at 200℃ for 30 minutes until done. Remove corn and unwrap the foil. Serve right away.

奶油烤白菜

Baked Cream Cabbage

烤箱用 烤/ 模具 cookware	烘烤功能 Functions	溫度°C	時間 min
耐熱烤皿 Heat Resistant Baking Utensils	▭	200	15

材料

奶油30公克、低筋麵粉30公克、牛奶300公克、白菜300公克、橄欖油1大匙、紅蔥頭末1大匙、披薩起司75公克、帕米森起司粉2大匙

❶ 鹽2小匙、黑胡椒粉1小匙、大蒜粉1小匙

做法

1. 將奶油放入鍋中，以小火加熱融化，倒入麵粉以網狀攪拌匙拌勻成麵糊，倒入牛奶繼續攪拌至濃稠且沸騰，加入材料❶調味後關火，即成奶焗醬。
2. 將白菜切大塊後汆燙至葉片軟化，撈起白菜放入烤皿，此時開啓烤箱預熱。
3. 把奶焗醬倒入烤皿，並淋上橄欖油和紅蔥頭末拌勻，表面撒上披薩起司，放入烤箱以220℃烘烤15分鐘，取出烤好的奶油白菜，撒上帕米森起司粉即可趁熱食用。

Tips　這道料理可以在製作完成後，放入鋁箔模型內待降溫，以保鮮膜包緊，放入冰箱冷凍，等到要食用前取出退冰，再放入烤箱烘烤即可。

Ingredients

30g butter, 30g cake flour, 300g milk, 300g napa cabbage, 1T. olive oil, 1T. fried minced shallots,75g pizza cheese, 2T. parmesan cheese powder

❶ 2t. salt, 1t. ground black pepper, 1t. garlic powder

Methods

1. To prepare the butter cream sauce: Melt butter in pan over low heat, stir in flour with a wiring mixing spoon and continue stirring to make batter. Pour in milk and stir until thick and boiling, then add ingredient ❶ to taste and remove from heat.
2. Cut napa cabbage into large pieces and blanch in boiling water until the leaves are softened. Remove napa cabbage to a baking bowl and preheat oven.
3. Pour the butter cream sauce in bowl and drizzle with olive oil and sprinkle with fried minced shallot. Mix well and sprinkle the surface with pizza cheese. Bake in oven at 220℃ for 15 minutes until done. Remove and serve right away while still steaming.

Tips

After this dish is prepared, it can be placed in an aluminum foil mold until the temperature drops, then wrapped up with saran wrap tightly and freeze in the freezer. Defrost when needed, and bake in oven before serving..

Baked Cream Cabbage

昆布蘿蔔煮

Konbu with Daikon

烤箱用 烤/ 模具 cookware	烘烤功能 Functions	溫度℃	時間 min
耐熱碗型烤皿 Heat Resistant Bowl-Shaped Baking Pans	or	200	60

材料

昆布1片(長10公分、寬3公分)、白蘿蔔300公克、豬五花肉125公克

❶ 醬油2大匙、醬油膏1大匙、黑糖1大匙、黑醋1小匙、清水350公克、米酒1大匙、味醂1小匙

做法

1. 將昆布洗淨放入烤皿。
2. 白蘿蔔去皮切塊與豬五花肉放入烤皿。此時開啓烤箱預熱。
3. 加入材料❶將所有材料拌勻，均勻沾上醬色。
4. 將烤皿放入烤箱，以200℃烘烤60分鐘，取出煮好的昆布，切成細條狀，即可趁熱食用。

Tips 可用梅花肉或後腿肉來取代五花肉，但是肉片不可太薄，以免在烘烤的過程中變得乾硬。

Ingredients

Konbu (10cm long 3cm wide), 300g daikon radish, 125g belly pork

❶ 2T. soy sauce, 1T. thick soy sauce, 1T. dark brown sugar, 1t. black vinegar, 350g water, 1T. rice wine, 1t. mirin

Methods

1. Rinse konbu well and place in a baking bowl.
2. Peel radish and cut into pieces, then place in the bowl along with belly pork. Preheat oven.
3. Add ingredient ❶ and mix well to make sure all the ingredients are evenly coated with soy sauce.
4. Place the bowl in the oven and bake at 200℃ for 60 minutes until done. Remove the konbu and cut into fine strips. Serve hot.

Tips

Belly pork can be substituted with marble pork or pork leg meat. However do not slice too thickly, or the texture of the meat will toughen during baking.

Konbu with Daikon

清燉元盅雞湯

Clear Stewed Chicken Soup in Pottery Cup

烤箱用 烤/ 模具 cookware	烘烤功能 Functions	溫度℃	時間 min
耐熱碗型烤皿 Heat Resistant Bowl-Shaped Baking Pans	or	220	50

材料

雞腿1支、嫩薑2片、蔥1支、清水750c.c.、鹽2小匙

做法

1. 將雞腿洗淨放入鍋中，加入份量外的清水煮沸一次，去除雜質和血水。
2. 撈起雞腿再次沖淨，放入耐熱器皿，加入嫩薑片、整支蔥和清水，蓋上鋁箔紙或是耐熱蓋。
3. 開啓烤箱預熱，將湯碗移入烤箱，以上下火200℃烘烤90分鐘。
4. 取出湯，打開鋁箔紙，撈除薑片和蔥，加鹽調味即可食用。

Tips　如果是使用厚的陶瓷器皿來盛裝雞湯，建議先將器皿浸泡在水中至少1小時，這樣可以確保熱氣順利穿透，讓湯更美味。

Ingredients

1 chicken leg, 2 slices young ginger, 1 scallion, 750c.c. water, 2t. salt

Methods

1. Rinse chicken leg and place in the pan, add water to cover the chicken and cook for some time to get rid of any impurities and blood.
2. Remove the chicken and rinse until clean, then transfer to a thick earthenware bowl with young ginger slices, whole scallion and water added. Cover with aluminum foil or cover.
3. Preheat oven and place bowl in the oven, cook with both elements at 200℃ for 90 minutes until done.
4. Remove the bowl and take off the foil, then discard ginger and scallion, season with salt to taste. Ready to serve.

Tips

If a thick earthenware utensil is used to hold the chicken soup, soaking the utensil in water at least 1 hour is suggested, to ensure the hot steaming air permeates it smoothly, so that the soup tastes better.

Clear Stewed Chicken Soup in Pottery Cup

清燉牛肉蘿蔔湯

Clear Stewed Beef with Daikon Soup

烤箱用 烤/模具 cookware	烘烤功能 Functions	溫度℃	時間 min
耐熱碗型烤皿 Heat Resistant Bowl-Shaped Baking Pans	or	220	90

材料

牛腩塊300公克、白蘿蔔300公克、清水1,500c.c.、嫩薑3片、鹽11/2大匙

做法

1. 將牛腩塊洗淨放入鍋中，加入份量外的清水煮沸一次，去除雜質和血水。
2. 撈起牛腩再次沖淨，放入耐熱器皿內，加入去皮切塊的蘿蔔、薑片和清水，蓋上鋁箔紙或是耐熱蓋。
3. 開啓烤箱預熱，將湯碗移入烤箱，以上下火220℃烘烤90分鐘。
4. 取出湯，打開鋁箔紙，撈除薑片，加鹽調味即可食用。

Tips 蘿蔔的皮較厚，如果僅使用刨刀無法完全將厚皮削下，因此去皮時，務必使用菜刀將皮刮下，否則蘿蔔不易煮熟。

Ingredients

300g beef finger ribs, 300g daikon radish, 1500c. c. water, 3 young ginger slices, 1 1/2T. salt

Methods

1. Rinse beef finger ribs well and place in a cooking pan with enough water to cover the beef. Cook until boiling to remove any impurities or blood.
2. Remove the beef and rinse well, then place in a thick china bowl along with peeled radish chunks, young ginger slices and water added. Cover with aluminum foil or cover.
3. Preheat the oven and place the bowl inside, then cook with both elements at 220℃ for 90 minutes until done.
4. Remove the soup and take off the foil, then discard the ginger slices and season with salt to taste. Ready to serve.

Tips

The skin of daikon radish is thick and it is difficult to remove with a peeler. Use a kitchen knife to remove the skin, or it will be difficult to cook the radish.

Clear Stewed Beef with Daikon Soup

南瓜酥皮濃湯

Pastry Over Pumpkin Cream Soup

烤箱用 烤/ 模具 cookware	烘烤功能 Functions	溫度℃	時間 min
耐熱碗型烤皿 Heat Resistant Bowl-Shaped Baking Pans	or	220	8～10

材料

冷凍酥皮4片、鹽1/2大匙、蛋液適量

❶ 南瓜泥300公克、紅蘿蔔70公克、無糖動物鮮奶油300公克、清水600c.c.、中筋麵粉30公克

❷ 馬鈴薯丁35公克

做法

1. 將材料❶放入果汁機攪拌均勻，接著倒入鍋中以小火加熱，加熱時必須不停攪拌，以免沾黏鍋底。
2. 加入材料❷再次煮至沸騰，即可關火，加鹽調味。此時開啓烤箱預熱。
3. 準備濃湯碗，在冷凍酥皮表面塗抹蛋液。
4. 將南瓜濃湯舀入碗中，每個碗蓋上一片酥皮，放入烤箱以220℃烘烤8至10分鐘。

Tips 從市場購回的南瓜，可以事先剖半去籽後放入電鍋蒸熟，將南瓜泥挖出放入容器或是保鮮袋，待完全降溫後移入冰箱冷凍，需要時再取出使用，非常方便。

Ingredients

4 frozen puff pastry sheets, 1/2T. salt, egg liquid as needed

❶ 300g mashed pumpkin, 70g carrot, 300g. unsweetened whipping cream, 600 c.c.water, 30g all purpose flour

❷ 35g potato dices

Methods

1. Blend ingredient ❶ in a blender until evenly mixed, pour into a pan and cook over low heat. Stir constantly during cooking to prevent from sticking to the bottom of the pan.
2. Add ingredient ❷ and cook until boiling, then remove from heat and season with salt to taste. Preheat the oven.
3. Preheat a soup bowl and spread the egg evenly across the frozen pastry sheets.
4. Scoop pumpkin soup into the bowl and top with a pastry sheet. Bake in oven at 220℃ for 8 to 10 minutes until done. Serve.

Tips

After the pumpkin is purchased from the market, it can be halved and seeded, then steamed until mushy and done in a rice cooker. Scoop out the mushy pumpkin flesh in a container or an air tight plastic bag, wait until cool, remove, and freeze. Remove portions as desired when needed.

Pastry Over Pumpkin Cream Soup

洋蔥濃湯 Onion Cream Soup

烤箱用 烤/ 模具 cookware	烘烤功能 Functions	溫度℃	時間 min
耐熱碗型烤皿 Heat Resistant Bowl-Shaped Baking Pans		230	5

材料

大顆洋蔥2顆、無鹽奶油50公克、橄欖油1大匙、低筋麵粉50公克、鮮雞高湯2,000c.c.
鹽2小匙、黑胡椒粉1/2小匙、法國麵包1條、塗抹用奶油40公克、乳酪絲100公克

做法

1. 洋蔥去皮切細絲，鍋中放入奶油和橄欖油加熱至奶油融化且呈淡褐色。此時放入洋蔥絲拌炒，炒至洋蔥變褐色，接著放入麵粉炒勻，最後倒入雞高湯轉中火煮15至20分鐘關火。
2. 濾出洋蔥湯，將洋蔥絲放入果汁機內快速攪打幾秒鐘（不需打成泥狀）。
3. 將打碎的洋蔥料放入鍋中與洋蔥湯混合，加熱沸騰後，以鹽及胡椒粉調味後關火。
4. 法國麵包切1公分厚片，抹上奶油，撒上乳酪絲，放入小烤箱烤至酥脆。
5. 食用前，將洋蔥湯倒入耐熱烤皿，表面擺上麵包片，再撒上少許的乳酪絲，放入烤箱以230℃焗烤5分鐘，即可出爐食用。

Tips 鮮雞高湯也可以改成蔬菜高湯或牛肉高湯，洋蔥炒至褐色需要花一點時間，請耐心拌炒。

Ingredients

2 large onions, 50g unsalted butter, 1T. olive oil, 50g cake flour, 2000c.c. fresh chicken stock, 2t. salt, 1/2t. black pepper powder, 1 French bread stick, 40g spray butter, 100g shredded cheesec

Methods

1. Remove skin from onions and shred finely. Melt butter and olive oil in pan until light brown, then stir-fry onion until brown, pour in chicken stock and turn the heat to medium. Cook for 15 to 20 minutes until done, then remove from heat.
2. Pour the onion soup through a sifter and blend the onion dregs in blender for a few second (no mashing necessary).
3. Add the blended onion back to the onion soup and cook in pan until boiling, then add salt and pepper to taste. Remove from heat.
4. Cut French bread stick into 1cm long slices, spread with spray butter and sprinkle with shredded cheese. Bake in small oven until crispy and crunchy.
5. Before serving, pour the onion cream soup into a heat resistant baking bowl, top with bread slices and sprinkle with shredded cheese. Cook in oven at 230℃ for 5 minutes. Remove and serve.

Tips

Fresh chicken stock can be substituted with vegetable soup stock or beef stock. It takes time to stir-fry the onions until brown, so just be patient.

Onion Cream Soup

Main Courses & Disbes

主食和菜餚

從豐腴鮮肉到營養菜飯，從經典歐洲料理到傳統中國菜，烤箱教你簡單做晚飯！

檸檬烤雞腿

Roasted Lemon Chicken Leg

烤箱用 烤/模具 cookware	烘烤功能 Functions	溫度℃	時間 min
耐熱烤皿 Heat Resistant Baking Utensils	or	200	20

材料

雞腿1隻、馬鈴薯1個、無鹽奶油100公克、檸檬1個

❶ 鹽1大匙、黑胡椒粉1小匙

做法

1. 雞腿洗淨，以廚房紙巾擦乾水分。
2. 將無鹽奶油放入攪拌盆中，置室溫下軟化。檸檬切半，擠出全部的汁。
3. 將檸檬汁慢慢與軟化的奶油混合，直到所有檸檬汁全部加入，即成檸檬奶油抹醬。
4. 把雞腿的皮撐開，取2至3大匙的抹醬塗抹在皮和肉中間，最後在表面撒上材料❶，並以綿線將雞腿綑綁固定。
5. 此時開啓烤箱預熱。馬鈴薯洗淨切成大塊狀，與雞腿一起放入烤箱，以200℃烘烤20分鐘。
6. 取出烤好的雞腿，用剪刀剪去綿線，將雞腿和馬鈴薯放置在盤中，馬鈴薯表面撒上份量外適量的鹽及胡椒，並搭配檸檬奶油食用即可。

Tips　檸檬可去腥味，也可以增加雞腿的香氣。食用前若再撒上少許的檸檬皮絲，擺盤將更漂亮。

Ingredients

1 chicken leg, 1 potato, 100g unsalted butter, 1lemon

❶ 1T. salt, 1t. ground black pepper

Methods

1. Rinse chicken leg and dry well with kitchen towel.
2. Soften the unsalted butter in a mixing bowl at room temperature. Cut lemon in halves and squeeze the juice out completely.
3. Combine lemon juice slowly with the softened butter, a little at a time until done, to make lemon butter spread.
4. Peel the skin of the chicken leg up to spread the lemon butter between skin and flesh, then sprinkle ingredient ❶ on the chicken leg surface.
5. Preheat oven. Rinse potato well and cut into large pieces, then place in oven along with the chicken leg. Roast at 200℃ for 20 minutes until done.
6. Remove the chicken and potato to a serving plate. Sprinkle potato pieces with salt and pepper as needed. Serve with lemon butter spread.

Tips

Lemon removes the odor and enhances the fragrance of chicken. Sprinkling with some lemon peels before serving will make the dish even more beautiful.

Roasted Lemon Chicken Leg

芥末烤雞胸

Mustard Roast Chicken Breast

烤箱用烤/模具 cookware	烘烤功能 Functions	溫度℃	時間 min
耐熱烤皿 Heat Resistant Baking Utensils		200	40、10

材料

去骨雞胸肉2個、綠葉生菜100公克、小蕃茄6個

❶ 法式芥末醬200公克、橄欖油2大匙、義大利綜合香料1大匙、蜂蜜2大匙

做法

1. 將雞胸肉洗淨擦乾，均勻塗上❶，醃漬30分鐘。
2. 此時開啟烤箱預熱，準備耐熱烤皿，將雞胸肉放入烤皿。
3. 以200℃烘烤40分鐘，拉出烤盤，將蜂蜜塗抹在雞胸肉表面，再放入烤箱續烤10分鐘。
4. 取出烤好的雞胸肉切片，搭配切絲的生菜和小蕃茄即可食用。

Tips 法式芥末醬含有芥末顆粒，平常可以搭配水煮肉腸、塗抹三明治、佐配燻雞肉或火雞肉。

Ingredients

2 boneless chicken breast, 100g green lettuce, 6 cherry tomatoes

❶ 200g French mustard sauce, 2T. olive oil, 1T. Italian spice mix, 2T. honey

Methods

1. Rinse chicken breast well and dry. Brush evenly Ingredients ❶ Let marinate for 30 minutes.
2. Preheat oven. Place chicken breast on a baking sheet.
3. Roast at 200℃ for 40 minutes. Remove baking sheet and brush honey evenly over the breast. Return to oven and continue roasting for 10 more minutes.
4. Remove chicken breast and cut into slices. Serve with shredded lettuce and cherry tomatoes on the side.

Tips

French mustard sauce contains mustard seeds. It can be served with boiled sausage, smoked chicken, turkey meat or as a sandwich spread.

Mustard Roast Chicken Breast

美式烤肋排

American Style Barbecue Ribs

烤箱用 烤/ 模具 cookware	烘烤功能 Functions	溫度℃	時間 min
耐熱烤皿 Heat Resistant Baking Utensils	or	200	50

材料

豬肋排900公克

❶ 醃料：大蒜5～6個、薑泥1/2大匙、辣椒1～2支、黑醋1小匙、蠔油1大匙、味醂1大匙、蜂蜜1大匙、蕃茄醬3大匙、橄欖油2大匙

❷ 桃李水果沙拉：加州桃李2個、紫洋蔥35公克、甜紅椒35公克、青蔥1支

❸ 檸檬汁1大匙、橄欖油1大匙、細砂糖1小匙、開心果粒11/2大匙

做法

1. 豬肋排以❶醃漬1小時。
2. 材料❷切小丁，與材料❸混合拌勻，即成桃李水果沙拉。
3. 此時開啟烤箱預熱。將醃漬好的豬肋排放入烤皿內，放入烤箱以200℃烘烤50分鐘，烘烤的過程中，必須用毛刷反覆將醃料塗抹在豬肋排表面約2至3次。
4. 將烤好的肋排取出切段，食用時搭配桃李水果沙拉即可。

Tips 購買豬肋排時，可請肉販將肋排肉片割下，只留下少許與肋排連接的肉，這樣可以縮短烘烤時間。

Ingredients

900g pork ribs

❶ Marinade: 5-6 cloves garlic, 1/2T. mashed ginger, 1-2 chili peppers, 1t. black vinegar, 1T. oyster sauce, 1T. mirin, 1T. honey, 3T. ketchup, 2T. olive oil

❷ Side Dish: 2 California peaches, 35g purple onion, 35g red bell peppers, 1 scallion

❸ 1T. lemon juice, 1T. olive oil, 1t. fine granulated sugar, 1 1/2T. eashews.

Methods

1. Marinate ribs in marinade❶ for 1 hour.
2. To prepare the side dish: Dice ingredient ❷ and combine well with ingredient ❸.
3. Preheat oven. Place ribs on the baking sheet and roast at 200℃ for 50 minutes. Brush the marinade at least 2 to 3 times on the ribs during roasting.
4. Remove ribs and cut into sections. Serve with side dish on the side.

Tips

When purchasing ribs, have the butcher slice the meat off the ribs, leaving only a little meat attached the ribs. This way the cooking time will be shortened.

American Style Barbecue Ribs

烤漢堡排 Roast Pattie

烤箱用 烤/模具 cookware	烘烤功能 Functions	溫度℃	時間 min
耐熱烤盤+鋁箔紙 Heat Resistant Baking Sheet+Aluminum Foil	or	220、175	10、25

材料

豬絞肉500公克、蛋1顆、麵包粉2大匙、洋蔥碎75公克、小馬鈴薯100公克、生菜100公克

❶ 鹽1 1/2小匙、黑胡椒粉1/2小匙、大蒜粉1小匙、橄欖油1小匙、義大利綜合香料1小匙

❷ 紅酒300c.c.、高湯150c.c.、蕃茄糊1大匙、黑醋1小匙、玉米粉1小匙

❸ 奶油30公克、鹽1小匙

做法

1. 將所有材料放入攪拌盆，加入材料❶，用筷子順同一個方向快速且用力攪拌5分鐘。
2. 攪拌完成的材料覆上保鮮膜，放入冰箱冷藏鬆弛30分鐘。
3. 此時開啟烤箱預熱，鋁箔紙上塗少許份量外的橄欖油。
4. 取出鬆弛完成的材料，分成每個125公克，放在手掌上左右拍打，以拍出裡面多餘的空氣，讓漢堡肉更紮實。
5. 將材料放置在鋁箔紙上，送入烤箱，先以220℃烘烤10分鐘，再降至175℃烘烤25分鐘。
6. 材料❷混合拌勻放入小鍋，以小火邊加熱邊攪拌，沸騰後加入材料❸拌勻即成淋醬。
7. 取出烤好的漢堡排放在盤中，淋上醬料並搭配蔬菜和水煮馬鈴薯即可。

Tips 絞肉最好瘦肥參半，這樣可以讓漢堡吃起來香甜有汁。淋醬的材料可以加入少許檸檬汁，讓醬汁略帶酸酸的味道，可以解膩開胃。

Ingredients

500g ground pork, 1 egg, 2T. bread crumb, 75g chopped onion,100g, small potatoes, 100g lelluce

❶ 1 1/2t. salt, 1/2t. ground black pepper, 1t. ground garlic, 1t. olive oil, 1t. Italian spice mix 300cc red wine, 150cc soup broth, 1T. tomato paste, 1t. black vinegar, 1t. corn flour

❷ 30g butter, 1t. salt

❸ 100g small potatoes, 100g lettuce

Methods

1. Combine all the ingredients in a mixing bowl, add ingredient ❶ and stir at the same direction rapidly with force for 5 minutes.
2. Wrap the method up with saran wrap and refrigerate for 30 minutes to relax.
3. Preheat oven. Brush the aluminum foil with a little olive oil.
4. Remove the pork and divide into portions of 125 grams each. Hold in hands and pat with palms hard enough to push out the extra air inside, so that the patties can be tight and firm.
5. Remove the patties to the aluminum foil and bake in oven at 220℃ for 10 minutes first, then reduce heat to 175℃ and continue baking for about 25 minutes.
6. Combine ingredient ❷ well in a saucepan. Cook over low heat while stirring until boiling, then add ingredient ❸ to mix. This is the sauce for drizzling over patties.
7. Remove the patties to the serving plate and drizzle with sauce. Serve with vegetables and potatoes as side dishes.

Tips

It is better to purchase half lean and half fat ground pork, as the pattie will taste sweet and juicy. A little lemon juice can be added to the drizzling sauce. The slightly sour sauce helps to stimulate the appetite and reduce greasiness.

Roast Pattie

迷迭香羊排

Rosemary Lamb Chops

烤箱用 烤/ 模具 cookware	烘烤功能 Functions	溫度℃	時間 min
烤箱網架＋鋁箔紙 Oven Baking Rack+Aluminum Foil	or	200	40

材料

羊排400公克、新鮮迷迭香葉2束、檸檬1個

❶ 醃料：橄欖油2大匙、鹽2小匙、黑胡椒粉1/2小匙

做法

1. 迷迭香葉以熱水洗淨（幫助迷迭香的香氣散出），將迷迭香葉切碎。
2. 羊排表面撒上材料❶和迷迭香葉，醃漬30分鐘。此時開啓烤箱預熱。
3. 將檸檬切片舖在鋁箔紙上，上面擺上醃漬完成的羊排，放入烤箱以200℃烘烤40分鐘，即可取出食用。

Tips　也可以使用整排未切的羊肩肉，以相同溫度烘烤90分鐘，烤好後再切片。另外，薄荷醬及迷迭香都和羊排的味道相搭。

Ingredients

400g lamb chops, 2 bunches fresh rosemary,1 lemon

❶ 2T. olive oil, 2t. salt, 1/2t. ground black pepper

Methods

1. Rinse rosemary well with hot water to help activate the aroma of the rosemary, then chop the leaves finely.
2. Sprinkle lamb chops evenly with ingredient ❶ and rosemary. Marinate for 30 minutes until the flavor is well absorbed. Preheat oven.
3. Cut lemon into thin slices and spread over aluminum foil. Place lamb chops on top of lemon slices and cook in oven at 200℃ for 40 minutes until done. Remove and serve.

Tips

The whole uncut lamb can be used in this recipe. Cook at the same temperature for 90 minutes until done, then slice. Mint sauce and rosemary are perfect with lamb chops.

Rosemary Lamb Chops

啤酒豬腳

Beer Sauce Marinated Pig Knuckle

烤箱用 烤/模具 cookware	烘烤功能 Functions	溫度℃	時間 min
耐熱烤皿 Heat Resistant Baking Utensils	or	200、150	15、150

材料

豬腳1個、清水500c.c.、啤酒375c.c.、蔥3支、大蒜3瓣

❶ 橄欖油1大匙、紅蔥頭3粒、鹽1小匙

❷ 檸檬1個、大蒜10瓣、青蒜2支、小蕃茄6個、橄欖油2大匙

做法

1. 將豬腳洗淨放入大鍋中，加入清水、啤酒、蔥和大蒜，煮至沸騰。
2. 轉小火續煮約30分鐘，關火，透過濾網濾出汁液。
3. 紅蔥頭切末，將材料❶混合均勻。此時開啓烤箱預熱。
4. 取出豬腳放在烤皿上，表面塗上做法3，淋上做法2的汁液放入烤箱以200℃烘烤15分鐘，再將烤溫降至150℃烘烤150分鐘。烘烤過程必須使用毛刷反覆將汁液塗抹在豬腳表面。
5. 將材料❷的檸檬切片、青蒜切小段，和其餘的材料❷混合均勻並淋上橄欖油，於最後30分鐘左右，把上述材料放在豬腳周圍，待烘烤完畢即可搭配食用。

Tips　市售的德國豬腳常以硝酸鹽當保色劑，因此肉質呈漂亮的粉紅色，而自製豬腳不使用化學添加物，雖然肉質不夠漂亮，卻保有健康。

Ingredients

1 pig knuckle, 500c.c. water, 375c.c. beer, 3 scallions, 3 cloves garlic

❶ 1T. olive oil, 3 cloves fried shallots, 1t. salt

❷ 1 lemon, 10 cloves garlic, 2 leeks, 6 cherry tomatoes, 2T. olive oil

Methods

1. Rinse knuckle well and cook in a big pan with water, beer, scallions and garlic cloves added until boiling vigorously.
2. Turn the heat down to low and continue cooking for 30 minutes, then remove from heat. Pour through a sifter to remove any dregs.
3. Mince shallots and combine ingredient ❶ well. Preheat oven.
4. Remove pig's foot to the baking sheet and brush the surface with method 2 . Drizzle with sauce from method 3 , then bake in oven at 200℃ for 15 minutes. Decrease the temperature to 150℃ and bake for 150℃ minutes. Repeat brushing the beersance over the pork repeatedly during baking.
5. Slice lemon thinly. Cut leeks into small sections. Combine all the ingredients ❷ well together and surround the pork 30 minutes ahead before it is done. Serve together.

Tips

Pig's feet sold in the market are all prepared nitrates as colorings, thus the meat appears to be beautiful pink. Homemade Pig's knuckle do not contain any artificial chemicals. Though the meat might not look as beautiful, it is healthier.

Beer Sance Marinaeed Pig Knuckle

椒鹽烤鴨

Pepper Salt Roast Duck

烤箱用 烤/ 模具 cookware	烘烤功能 Functions	溫度℃	時間 min
烤箱旋轉棒＋烤盤 Oven Rotisserie Skewer+ Heat Resistant Baking Sheet	or	130、200	7、50

材料

鴨1隻、鹽70公克、花椒粒35公克

做法

1. 將花椒粒放入烤箱，以130℃烘烤7分鐘。
2. 取出烤好的花椒粒，混合鹽放入研磨缽或研磨機，研磨成細粒狀，即成花椒鹽。
3. 取適量花椒鹽平均塗抹在鴨子表面。此時開啓烤箱預熱。
4. 鴨子以烤箱的旋轉棒固定後放入烤箱，以200℃烘烤50分鐘。
5. 把烤好的鴨子取出，用刀子將鴨肉片下，即可食用。或是搭配甜麵醬、青蒜絲和荷葉餅捲起後食用。

Tips 使用旋轉架烘烤整隻家禽時，務必在烤箱底部放置一個有水的烤盤，這樣可以承接在烘烤過程中滴下的油。

Ingredients

1 duck,70g salt, 35g peppercorns

Methods

1. Roast peppercorns in oven at 130℃ for 7 minutes.
2. Remove peppercorns to a mortar and pestle or a grinder, add salt and grind until fine to make peppersalt.
3. Spread peppersalt evenly over duck. Preheat oven.
4. Use a rotisserie skewer to secure duck in oven and roast at 200℃ for 50 minutes until done.
5. Remove duck from oven. Use a knife and slice away the duck meat. Serve with sweet bean paste, shredded leek and crepes for wrapping.

Tips

When using a rotisserie oven with a rotisserie skewer to roast the whole poultry, a baking tray full of water must be placed underneath to catch the dripping oil when roasting.

Pepper Salt Roast Duck

烤箱用 烤/ 模具 cookware	烘烤功能 Functions	溫度°C	時間 min
耐熱烤皿＋鋁箔紙 Heat Resistant Baking Utensils+ Aluminum Foil	 or	200	15、5

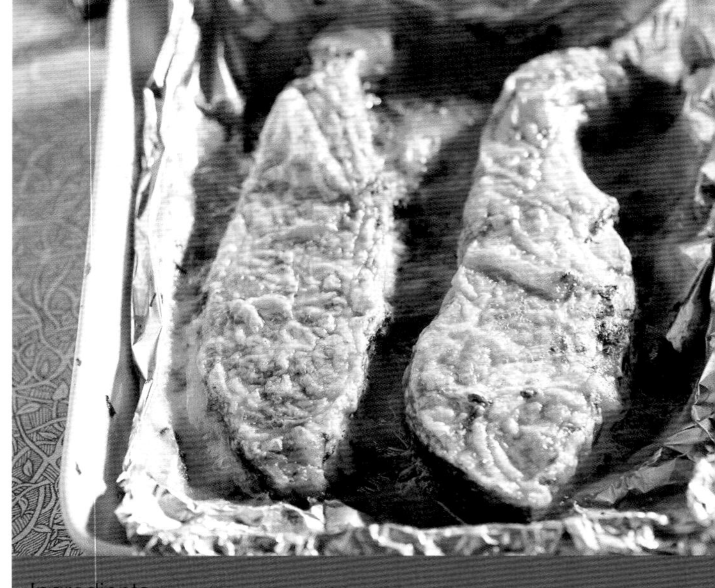

菠菜奶汁鮭魚排

Creamy Salmon Fillet with Spinach

材料

菠菜1/2把(約20公克)、鮭魚400公克、橄欖油1大匙、鹽1小匙

❶ 奶油30公克、低筋麵粉30公克、牛奶300公克

❷ 肉荳蔻粉1/4小匙、鹽1小匙

做法

1. 將材料❶的奶油放入鍋中以小火加熱融化，倒入麵粉以網狀攪拌匙拌勻成麵糊，倒入牛奶繼續攪拌至濃稠且沸騰，加入材料❷調味後關火，即成奶焗醬。
2. 菠菜切碎放入攪拌盆，混合橄欖油和鹽拌勻。把鮭魚放在鋁箔紙上，兩面均勻抹上菠菜料。此時開啓烤箱預熱。
3. 將鮭魚放入烤箱以200°C烘烤15分鐘；拉出烤盤，在鮭魚表面淋上2大匙的奶焗醬，再放入烤箱續烤5分鐘即可取出。

Tips　茴香和菠菜都與鮭魚非常搭配，也可以改用便宜又好吃的茴香來取代菠菜。

Ingredients

1/2 bunch spinach (approximately 20g), 400g salmon, 1T. olive oil, 1t. salt

❶ 30g butter, 30g cake flour, 300g milk

❷ 1/4t. nutmeg powder, 1t. salt

Methods

1. To prepare the cream sauce: Melt butter from ingredient ❶ in pan over low heat, add flour and stir with spatula evenly to make batter, then pour in milk and stir until the batter is thick and boiling. Add ingredient ❷ and remove from heat.
2. Chop spinach finely and remove to a mixing bowl along with olive oil and salt added, then stir until evenly mixed. Place salmon over the aluminum foil and spread both sides evenly with spinach filling. Preheat oven.
3. Bake salmon in oven at 200°C for 15 minutes. Remove baking sheet and drizzle the salmon with 2 tablespoons of cream sauce, then return to oven and continue baking for 5 minutes longer.

Tips

Both fennel leaves and spinach are perfect for salmon, thus spinach can be substituted with cheaper and more delicious fennel leaves.

烤箱用 烤/ 模具 cookware	烘烤功能 Functions	溫度°C	時間 min
耐熱烤皿＋鋁箔紙 Heat Resistant Baking Utensils+ Aluminum Foil	or	200	30

起司烤鱈魚

Baked Cod with Cheese

材料

鱈魚300公克、橄欖油1大匙、鹽1小匙、黑醋1小匙、無鹽奶油55公克

❶ 麵包粉1 1/2杯、披薩起司120公克、帕米森起司粉1大匙、青蔥末2大匙、義大利綜合香料2大匙

做法

1. 將橄欖油、鹽和黑醋在小碗中混合拌勻，淋在鱈魚表面，醃漬約30分鐘。
2. 把材料❶放在另一個乾淨的盆中，混合拌勻。此時開啓烤箱預熱。
3. 把醃漬完成的鱈魚放在鋁箔紙上，表面覆蓋材料❶，並均勻撒上奶油塊。
4. 鋁箔紙不需要包裹，直接把材料放入烤箱，以200℃烘烤30分鐘。

Tips　麵包粉不只可用來做油炸的材料，用來製作烤箱菜也非常適合，口感酥脆不油膩。

Ingredients

300g cod, 1T. olive oil, 1t. salt, 1t. black vinegar,55g unsalted butter

❶ 1 1/2C. bread crumbs, 120g pizza cheese, 1T. parmesan cheese, 2T. minced scallions, 2T. Italian spice combination

Methods

1. Combine olive oil, salt and black vinegar in a small bowl well, then drizzle over cod and marinate for about 30 minutes.
2. Combine ingredient ❶ well in another bowl. Preheat oven.
3. Place cod on an aluminum foil sheet, top with ingredient and sprinkle evenly with butter chunks.
4. There is no need to wrap up in the foil, just bake directly in oven at 200℃ for 30 minutes until done. Remove and serve.

Tips

Breadcrumbs not only can be used in deep-frying, but also in oven dishes. The resulting texture will be crispy and not at all oily.

烤箱用 烤/ 模具 cookware	烘烤功能 Functions	溫度℃	時間 min
耐熱烤皿＋鋁箔紙 Heat Resistant Baking Utensils+ Aluminum Foil	or	180	25

紙包洋蔥鮮鱸魚
Perch with Onion Wrapped in Paper

材料

鱸魚腹2片、紫洋蔥115公克、月桂葉1片、迷迭香1束、檸檬1顆

❶ 鹽2小匙、黑胡椒粉1/2小匙、大蒜粉1/2小匙、橄欖油1大匙

做法

1. 鱸魚腹洗淨擦乾，兩面均勻塗抹材料❶。
2. 洋蔥和檸檬切片，準備一張鋁箔紙或耐熱烤紙，將鱸魚放在紙上，舖上洋蔥、月桂葉、迷迭香和檸檬，將鋁箔紙包緊。此時開啓烤箱預熱。
3. 將材料放入烤箱，以180℃烘烤25分鐘。

Tips 油紙耐高溫適合烘烤，可以代替鋁箔紙，餐具材料行買得到。

Ingredients

2 perch stomach sections, 115g purple onion, 1 bay leaf, 1 bunch rosemary, 1 lemon

❶ 2t. salt, 1/2t. black pepper, 1/2t. garlic powder, 1T. olive oil

Methods

1. Rinse perch stomach and dry well, spread ingredient ❶ evenly on both sides.
2. Cut onion and lemon into slices. Prepare a sheet of aluminum foil or heat resistant parchment paper. Place perch stomach on top and spread onion, bay leaf, rosemary and lemon slices evenly on top. Wrap up tightly and preheat oven.
3. Bake in oven at 180℃ for 25 minutes until done. Remove and serve.

Tips

Parchment paper, which can be purchased at the local cooking supply store, is suitable for baking. It can be used as a substitute for aluminum foil.

烤箱用 烤/ 模具 cookware	烘烤功能 Functions	溫度°C	時間 min
耐熱烤皿 Heat Resistant Baking Utensils	or	180	20

清蒸蛤蜊絲瓜
Steamed Clams and Silk Squash

材料

蛤蜊300公克、絲瓜1個、鹽1小匙

❶ 薑片2片、橄欖油1大匙、米酒1小匙

做法

1. 絲瓜去皮後，切成四段，用菜刀將中心部份切除，只留厚實的瓜肉，切片。
2. 將絲瓜與蛤蜊放入攪拌盆，加入材料❶拌勻。此時開啓烤箱預熱。
3. 準備一個淺的耐熱烤皿，將材料放入烤皿，表面覆上鋁箔紙，放入烤箱以180℃烘烤20分鐘。
4. 取出完成的材料，把薑片丟掉，撒上鹽調味即可。

Tips　可以改用新鮮干貝或扇貝取代蛤蜊，材料中亦可加入1小匙枸杞。

Ingredients

300g clams, 1 silk squash

❶ 2 slices ginger, 1T. olive oil, 1t. rice wine, 1t. salt

Methods

1. Peel silk squash and cut into 4 equal sections, then remove the inner seed sections and retain the thick flesh.
2. Combine clams and silk squash in a mixing bowl along with ingredient ❶ added, then mix well. Preheat oven.
3. Preheat a shallow baking bowl and place the ingredients inside. Cover with aluminum foil and bake in oven at 180℃ for 20 minutes until done.
4. Remove and discard ginger slices, then sprinkle salt to taste. Serve.

Tips

Clams can be substituted with fresh scallops or Japanese abalone. 1 teaspoonful of lycium berries can be added to the ingredients.

烤甜椒明蝦沙拉

Baked Shrimp Salad with Bell Pepper

烤箱用 烤/ 模具 cookware	烘烤功能 Functions	溫度°C	時間 min
耐熱烤盤 Heat Resistant Baking Sheet	or	200	15

材料

明蝦400公克、大蒜5瓣、紅甜椒1/2個、黃甜椒1/2個、小蕃茄8顆、小黃瓜1條、青蔥2支、檸檬1顆

❶ 白酒50c.c.、橄欖油2大匙、鹽1小匙

做法

1. 明蝦去殼（保留尾部），大蒜整顆去皮。
2. 甜椒切小片、小蕃茄切半、小黃瓜切滾刀塊、青蔥的白色部份切碎，綠色部份則保留備用。
3. 把明蝦、大蒜和做法❷的材料放在鋁箔紙上，淋上材料❶拌勻。此時開啓烤箱預熱。
4. 將鋁箔紙包緊，放入烤箱以200℃烘烤15分鐘。
5. 將烤好的材料打開，移入盤中，檸檬切角狀放在盤邊，青蔥綠色的部份切絲，裝飾在沙拉上面即可。

Tips 青蔥可改用九層塔，亦可加入少許香菜提味。

Ingredients

400g prawns, 5 cloves garlic, 1/2 red bell pepper, 8 cherry tomatoes, 1 Chinese cucumber, 2 scallions,1 lemon

❶ 50c.c. white wine, 2T. olive oil, 1t. salt

Methods

1. Remove shells from prawns, but retain the tail sections. Remove skin from garlic cloves.
2. Cut bell pepper into small pieces. Halve cherry tomatoes. Roll cut Chinese cucumber into random pieces. Chop whites of scallions finely and retain the greens.
3. Place shrimp, garlic and ingredients from method on the aluminum foil sheet, drizzle with ingredient ❶ and mix well. Preheat oven.
4. Wrap up the shrimp well and bake in oven at 200℃ for 15 minutes.
5. Remove from oven and transfer to a serving plate. Cut lemon into wedges and garnish on the side. Shred green parts of scallions and place on top as garnish. Serve.

Tips

Scallions can be substituted with basil. Cilantro can be added to enhance the flavor.

Baked Shrimp Salad with Bell Pepper

茄汁鮮肉高麗菜卷

Pork Stuffed Cabbage Roll in Tomato Sauce

烤箱用 烤/ 模具 cookware	烘烤功能 Functions	溫度°C	時間 min
耐熱烤皿 Heat Resistant Baking Utensils	or	185	30

材料

高麗菜1顆

❶ 蕃茄糊2大匙、罐頭蕃茄丁400公克、豬骨高湯50c.c.、白酒3大匙、鹽1小匙、細糖2小匙、黑胡椒粉1/4小匙、大蒜粉1/2小匙

❷ 豬絞肉200公克、洋蔥35公克、太白粉1小匙、蛋白1顆

做法

1. 材料❶放入攪拌盆混合均勻，即成茄醬。
2. 材料❷放入另一個乾淨的攪拌盆，用筷子順著同一方向用力攪拌3分鐘，即成餡料。
3. 用菜刀將高麗菜的梗挖掉，整顆放入滾水中汆燙，燙至葉片軟化後取出高麗菜，用手將葉片剝下，葉片中心硬梗的部份切薄，以方便包裹。
4. 準備一個淺的耐熱烤皿，淋上份量外1大匙的橄欖油。
5. 取一片高麗菜打開，包入2大匙的內餡捲起，排入烤皿。重複此一動作直到所有的內餡使用完畢。
6. 此時開啓烤箱預熱，將茄醬淋在高麗菜卷上，放入烤箱以180℃烘烤30分鐘。

Tips　細長的山東白菜也很適合用來包裹菜卷，絞肉餡亦可改成魚漿或蟹肉。

Ingredients

1 head cabbage

❶ 2T. tomato paste, 400g diced canned tomato, 50c.c. pork bone soup broth, 3T. white wine, 1t. salt, 2t. fine granulated sugar, 1/4t. black pepper, 1/2t. garlic powder

❷ 200g ground pork, 35g onion, 1t. cornstarch, 1 egg white

Methods

1. To prepare the tomato sauce: Combine ingredient ❶ in a mixing bowl well together.
2. Put ingredient ❷ in another mixing bowl, stir at the same direction with chopsticks vigorously for 3 minutes to form filling.
3. Cut off the stem from the head of cabbage with a kitchen knife, then blanch in boiling water without tearing off the leaves. Blanch until the leaves are softened, then remove from water. Peel off leaves one by one with hands and trim the center stem more thinly, so that it is easier to wrap it.
4. Prepare a shallow heavy-duty heat resistant baking sheet and drizzle with 1 tablespoon of olive oil (not listed in the ingredients).
5. Spread one cabbage leaf wide open, place 2 tablespoons of filling in center, then wrap up and roll into a cylinder. Arrange in baking utensil. Repeat this process until all the rolls are done.
6. Preheat oven. Drizzle tomato sauce over cabbage rolls and cook in oven at 180℃ for 30 minutes.

Tips

Long su choy cabbage is also suitable for wrapping up the filling. Ground pork filling can be substituted with fish paree or crab meat.

Pork Stuffed Cabbage Roll in Tomato Sauce

培根里肌卷

Bacon and Tenderloin Rolls

烤箱用 烤/ 模具 cookware	烘烤功能 Functions	溫度℃	時間 min
耐熱烤皿 Heat Resistant Baking Utensils	or	185	45

材料

豬里肌300公克、培根6條、洋蔥75公克、玉米筍150公克、四季豆100公克

❶ 五香粉1小匙、鹽1/2小匙、米酒1大匙

❷ 紅酒100c.c.、黑醋2小匙、蜂蜜1大匙

做法

1. 豬里肌表面以材料❶ 醃漬1小時；四季豆去蒂頭並撕除硬鬚。紅酒100c.c.、黑醋2小匙、蜂蜜1大匙
2. 準備耐熱烤皿，將材料❷倒入烤皿拌勻。洋蔥切絲撒在上面。
3. 豬里肌切6小段，包入切小段的玉米筍或四季豆，再以培根包裹擺入烤皿。
4. 此時開啓烤箱預熱，放入材料以185℃烘烤45分鐘。

Tips 豬里肌片在包捲前，可以先用槌肉棒拍打，讓肉質鬆弛，更有利於包捲。

Ingredients

300g pork tenderloin, 6 strips bacon, 75g onion, 150g baby corn, 100g string beans

❶ 1t. five spice powder, 1/2t. salt, 1T. rice wine

❷ 100cc red wine, 2t. black vinegar, 1T.honey

Methods

1. Marinate pork tenderloin in ingredient ❶ for 1 hour. Remove both ends from green beans and tear off the strings.
2. Prepare a heat resistant baking bowl, mix ingredient ❷ in bowl until even. Shred onion and sprinkle on top.
3. Cut pork tenderloin into 6 small sections. Take a piece of tenderloin and wrap baby corn sections or green bean sections up first. Spread a bacon strip wide and place tenderloin roll in center. Roll up tightly to form a cylinder. Repeat to finish all the rolls, then transfer them to the baking bowl and place in oven.
4. Preheat oven and bake at 185℃ for 45 minutes until done. Remove and serve.

Tips

Before wrapping, tenderize the pork by patting with a kicthen hammer, so that the meat can be more easily used for wrapping.

Bacon and Tenderloin Rolls

義式通心焗麵

Italian Style Macaroni and Cheese

烤箱用 烤/模具 cookware	烘烤功能 Functions	溫度°C	時間 min
耐熱烤皿 Heat Resistant Baking Utensils	or	220	10

材料

奶油30公克、低筋麵粉30公克、牛奶300公克、義大利通心麵200公克、大蒜2瓣、橄欖油1/2小匙、三色蔬菜110公克、培根75公克、洋蔥35公克、披薩起司75公克、帕米森起司粉1大匙、鹽1小匙、黑胡椒粉1/2小匙

做法

1. 奶油放入鍋中以小火加熱融化，倒入麵粉以網狀攪拌匙拌勻成麵糊，倒入牛奶繼續攪拌至濃稠且沸騰後關火，即成奶焗醬。
2. 煮一鍋水，加入大蒜和少許的橄欖油，沸騰後放入通心麵，煮約8分鐘，確認通心麵煮軟了之後，瀝出通心麵，大蒜丟棄不用。
3. 培根和洋蔥切丁，放入耐熱烤皿，倒入煮過的通心麵、三色蔬菜，並加鹽和胡椒粉調味拌勻。
4. 淋上奶焗醬，撒上披薩起司。此時開啓烤箱預熱。
5. 將材料放入烤箱烘烤，以220℃烘烤10分鐘，取出烤好的通心麵，表面撒上帕米森起司粉即可食用。

Tips　義大利麵不宜久煮，以免麵質過於軟爛而缺乏彈牙嚼感。

Ingredients

30g butter, 30g cake flour, 300g milk, 200g elbow, shell or penne macaroni, 2 cloves garlic, 1/2t. olive oil, 110g three colored vegetables, 75g bacon, 35g onion, 75g pizza cheese, 1T. ground parmesan cheese, 1t. salt, 1/2t. black pepper

Methods

1. Melt butter in pan over low heat, then fold in flour and stir with a wire mixing spoon until batter is formed. Add milk and continue stirring until thick and boiling, remove from heat.
2. Bring a pot of water to a boil, add garlic and a little olive oil, then bring to a boil and add macaroni. Cook for about 8 minutes, make sure the macaroni is soft, drain and discard garlic.
3. Dice bacon and onion, place in a heat resistant baking bowl, add cooked macaroni, three-colored vegetables, as well salt and pepper to taste. Mix well until even.
4. Drizzle with cream sauce and sprinkle with pizza cheese. Preheat oven. Place the macaroni and cheese in the oven.
5. Cook at 220℃ for 10 minutes. Remove and sprinkle with parmesan cheese and serve.

Tips

Do not cook Italian noodles for a long time, or they will turn out to be too soft and lose their chewy texture.

Italian Style Macaroni and Cheese

烤箱用 烤/模具 cookware	烘烤功能 Functions	溫度°C	時間 min
耐熱烤盤 Heat Resistant Baking Sheet	or or	220	15

臘腸海鮮披薩

Salami & Seafood Pizza

材料

披薩起司110公克

❶ 高筋麵粉300公克、冰水180公克、鹽1/2小匙、沙拉油1小匙、即溶酵母粉1小匙

❷ 市售披薩抹醬150公克、臘腸75公克(切薄片)、草蝦110公克、新鮮干貝75公克

做法

1. 將材料❶放入攪拌盆，搓揉成表面光滑的麵糰；麵糰以保鮮膜包裹，放置在室溫下鬆弛30分鐘。
2. 把鬆弛後的麵皮放置在撒有少量高筋麵粉的桌面上，用擀麵棍擀成直徑約24公分的薄片。準備一張烘焙紙或鋁箔紙，塗上少許橄欖油，將麵皮放在上面，蓋上保鮮膜或是擰乾的濕布，以免麵皮風乾變硬。
3. 此時開啓烤箱預熱。將披薩抹醬塗抹在麵皮上面，再將其餘的材料❷平均排列上去，最後撒上披薩起司，放入烤箱以220℃烘烤15分鐘即可。

Ingredients

110g pizza cheese

❶ 300g bread flour, 180g ice water, 1/2t. salt, 1t cooking oil, 1t. instant yeast

❷ 150g market-sold pizza sauce, 75g salami (slice thinly), 110g grass shrimp, 75g fresh scallops

Methods

1. Place ingredient ❶ in mixing bowl, stir, knead and roll until the dough is smooth and soft. Wrap up with saran wrap and let rise at room temperature for 30 minutes.
2. Place the risen dough on the floured working table, knead and roll with a rolling pin to make the dough into a 24cm in centimeter round thin pizza dough. Prepare a sheet of parchment paper or aluminum foil, brush with a little olive oil, then place the pizza dough on top. Cover with saran wrap or damp kitchen cloth to prevent the dough from drying out and hardening.
3. Preheat oven. Brush the pizza sauce over the dough and spread the ingredients ❷ evenly on dough. Sprinkle pizza cheese evenly over. Bake in oven at 220℃ for 15 minutes until done. Serve.

烤箱用 烤/ 模具 cookware	烘烤功能 Functions	溫度℃	時間 min
耐熱烤盤 Heat Resistant Baking Sheet	or or	220	15

夏威夷青蔬總匯披薩

Hawaii Green Combination Pizza

材料

披薩起司110公克

❶ 高筋麵粉300公克、冰水180公克、鹽1/2小匙、沙拉油1小匙、即溶酵母粉1小匙

❷ 市售披薩抹醬150公克、鳳梨片100公克、火腿125公克、青椒1個

做法

1. 將材料❶放入攪拌盆，搓揉成表面光滑的麵糰。麵糰以保鮮膜包裹，放置在室溫下鬆弛30分鐘。
2. 將鬆弛後的麵皮放在撒有少量高筋麵粉的桌面上，用麵棍擀成直徑約24公分的薄片，準備一張烘焙紙或鋁箔紙，塗上少許橄欖油，將麵皮放在上面，蓋上保鮮膜或擰乾的濕布，以免麵皮風乾變硬。
3. 此時開啓烤箱預熱。將抹醬塗抹在麵皮上面，再將其餘的材料❷平均排列上去，最後撒上披薩起司，放入烤箱以220℃烘烤15分鐘。

Ingredients

110g pizza cheese

❶ 300g cake flour, 180g ice water, 1/2t. salt, 1t cooking oil, 1t. instant yeast

❷ 150g market-sold pizza sauce, 100g pineapple slices, 125g ham, 1 green pepper

Methods

1. Place ingredient ❶ in mixing bowl, stir, knead and roll until the dough is smooth and soft. Wrap up with saran wrap and rise at room temperature for 30 minutes.
2. Spread the inner pan with butter, place the risen dough on a cake floured working table, roll out the dough with a rolling pin into a 24cm centimeter. Prepare a sheet of parchment paper or aluminum foil, brush with a little olive oil, then place the pizza dough on top. Cover with saran wrap or damp kitchen cloth to prevent the dough from drying out and hardening.
3. Preheat oven. Brush the pizza sauce over the dough and spread the ingredients ❷ evenly on dough. Sprinkle pizza cheese evenly over. Bake in oven at 220℃ for 15 minutes until done. Serve.

西班牙海鮮飯

Spanish Seafood Rice

烤箱用 烤/ 模具 cookware	烘烤功能 Functions	溫度℃	時間 min
耐熱烤皿＋鋁箔紙 Heat Resistant Baking Utensils+ Aluminum Foil	or	200	40

材料

白米2杯、清水2杯、白酒2大匙、鹽2小匙、黑胡椒粉1/2小匙

❶ 淡菜350公克、白肉魚100公克、草蝦125公克、蕃茄100公克、洋蔥35公克

❷ 紅蔥頭5個、橄欖油1大匙、番紅花1小匙或薑黃粉1大匙、月桂葉1片、薑片1片

做法

1. 所有海鮮洗淨瀝乾，蕃茄去皮去籽切丁，洋蔥也切丁，紅蔥頭切細末。
2. 白米以份量外的清水洗淨，瀝乾水分後，放入耐熱烤皿或陶鍋內舖平，倒入清水和白酒，放入材料❷，以飯匙拌勻，再舖上材料❶。
3. 烤皿上面蓋上鋁箔紙或耐熱鍋蓋，放入烤箱以200℃烘烤40分鐘。
4. 取出月桂葉和薑片，加入鹽和黑胡椒粉拌勻後即可食用。

Tips 西班牙海鮮飯偏好使用扁平的鍋子來製作，因為扁平的容器受熱均勻、能讓材料在短時間內煮熟。

Ingredients

2C. white rice, 2C. water, 2T. white wine,2t. salt, 1/2t. ground black pepper

❶ 350g mussels, 100g white fish meat, 125g grass shrimp, 100g tomato, 35g onion

❷ 5 shallot cloves, 1T. olive oil, 1t. saffron or 1T. turmeric powder, 1 bay leaf, 1 slice ginger

Methods

1. Rinse all the seafood and drain. Peel the skin off the tomatoes and discard seeds, then dice. Dice onion. Mince shallots finely.
2. Rinse rice well with water until clean, drain and spread evenly across a heat resistant baking bowl or crock pot. Pour in water and white wine along with ingredient ❷ added. Stir well with a spoon and then spread ingredient ❶ evenly over the mix.
3. Cover the bowl with aluminum foil or cover with a lid. Cook in oven at 200℃ for 40 minutes.
4. Remove the bowl, open the foil or lid, discard bay leaf and ginger slice, then add salt and black peper. Mix well and serve.

Tips

It is better to prepare this Spanish style seafood rice in a frying pan, because the flat pan produces heat more evenly and cooks the ingredients in a short time.

Spanish Seafood Rice

烤箱用 烤/ 模具 cookware	烘烤功能 Functions	溫度℃	時間 min
耐熱烤皿＋鋁箔紙 Heat Resistant Baking Utensils+ Aluminum Foil	or	150	50

港式臘腸煲飯

Hong Kong Style Sausage with Steamed Rice

材料

肝腸75公克、臘腸75公克、香腸75公克、白米2杯、清水21/4杯、青江菜110公克

做法

1. 白米以份量外的清水洗淨，瀝乾水份後，放入耐熱烤皿或陶鍋內舖平，再倒入清水。
2. 將肝腸、臘腸和香腸舖在白米表面，此時開啓烤箱預熱。
3. 烤皿以鋁箔紙或耐熱鍋蓋蓋上，放入烤箱以150℃烘烤50分鐘。
4. 取出烤好的三種腸，斜切成薄片；白飯以飯匙拌勻，再舖上切片的腸，最後搭配汆燙過的青菜食用即可。

Tips 使用烤箱煮飯，必須使用比電子鍋還要多一點的水分，因為烤箱的高溫容易把水份烘乾。在煮之前，如果能把米飯浸泡10～15分鐘，烤出來的米飯也會更好吃。

Ingredients

375g liver sausages, 75g lag chong Chinese dried sausages, 75g Chinese sausage, 2C. white rice, 2 1/4C. water, 110g baby bok choy

Methods

1. Rinse rice with water until clean and drain. Spread evenly in a heat resistant bowl or a crock pot, then add water.
2. Spread liver sausages, Chinese dried sausages and Chinese sausages across rice and preheat oven.
3. Cover the bowl with aluminum foil and a lid, place in oven and cook at 150℃ for 50 minutes until done.
4. Remove the bowl, cut the sausages diagonally into thin slices. Stir rice well with a spoon. Spread the sausage slices on top and serve with blanched baby bok choy on the side.

Tips

When using the oven to cook the rice, more water has to be added because the water evaporates faster in a high temperature oven than a rice cooker. Before baking, soaking rice in water for 10 to 15 minutes will make the rice taste even more delicious.

烤箱用 烤/ 模具 cookware	烘烤功能 Functions	溫度°C	時間 min
耐熱烤皿＋鋁箔紙 Heat Resistant Baking Utensils+ Aluminum Foil		220、130	10、90

港式五花叉燒

Hong Kong Style Barbecue Belly Pork

材料

豬五花肉600公克、青江菜100公克

❶ 蔥2支、薑3片、醬油5大匙、米酒2大匙、細砂糖2大匙、蕃茄醬1小匙、蜜汁烤肉醬2小匙、沙拉油1小匙

做法

1. 將材料❶混合放入攪拌盆，五花肉表面刻出刀痕，放入盆中醃漬4小時。其間必須翻面一次。
2. 開啓烤箱預熱。烤盤舖一張鋁箔紙，刷上少許份量外的沙拉油。
3. 將醃漬過的五花肉放置在烤盤上，放入烤箱以220℃烘烤10分鐘。取出五花肉，在表面刷上醃料，將烤箱降至130℃，放入五花肉續烤90分鐘，至肉完全烤熟即可。
4. 食用時，搭配汆燙過的青菜即可。

Tips 如果不喜歡叉燒偏甜的口感，可以省略掉配方中的細砂糖。

Ingredients

600g streaky pork 100g baby bok choy

❶ 2 scallions, 3 slices ginger, 5T. soy sauce, 2T. rice wine, 2T. fine granulated sugar, 1t. ketchup, 2t. honey flavored barbecue paste, 1t. cooking oil

Methods

1. Combine ingredient ❶ all together in a mixing bowl. Score streaky pork on the surface a few times and marinate for 4 hours. Turn the pork over during marinating.
2. SPreheat oven. Spread baking sheet with a piece of aluminum foil and brush with a little cooking oil.
3. Place streaky pork on the sheet and bake in oven at 220℃ for 10 minutes until done. Remove straky pork and brush marinade on the surface of the pork. Reduce the heat to 130℃ and return pork to oven. Continue roasting for 90 minutes longer until the meat is completely done.
4. Serve with blanched vegetables on the side.

Tips

If sweet flavor is not preferred, omit the fine granulated sugar in the ingredients.

透抽蔬菜咖哩飯

Squid Veggie Curry Rice

烤箱用 烤/模具 cookware	烘烤功能 Functions	溫度°C	時間 min
耐熱烤皿＋鋁箔紙 Heat Resistant Baking Utensils+ Aluminum Foil	or	200	50

材料

白米2杯、清水2 1/4杯、咖哩粉1大匙、透抽500公克(4條)、洋蔥75公克、紅蘿蔔35公克、青豆仁35公克、玉米粒50公克、鹽2小匙、黑胡椒粉1小匙

做法

1. 白米以份量外的清水洗淨，瀝乾水份後，放入耐熱烤皿或陶鍋內舖平，倒入清水和咖哩粉。
2. 透抽洗淨清除腸泥和皮膜。將洋蔥切丁，把洋蔥撒在米飯上，透抽也放在米飯上。
3. 烤皿蓋上鋁箔紙或耐熱鍋蓋，放入烤箱以200℃烘烤50分鐘。
4. 將紅蘿蔔、青豆仁和玉米粒放入沸水中，加少許鹽汆燙，撈起瀝乾。
5. 烤好的咖哩飯取出，將透抽夾起放在乾淨的盤子上，其餘材料混合三色蔬菜，用飯匙翻拌均勻，再加入少許鹽和胡椒粉調味。
6. 把拌好的米飯非常紮實的填入透抽裡。
7. 將透抽咖哩飯切成片狀，排列在盤上即可食用。

Tips 另一個簡易做法，是將透抽切成圈狀，與米粒及洋蔥混合後，放入烤箱加熱，可以省去把米飯填入透抽裡面的手續。

Ingredients

2C. white rice, 2 1/4C. water, 1T. curry powder, 500g squids (4 squids), 75g onion, 35g carrot, 35g peas, 50g corn kernels, 2t. salt, 1t. ground black pepper

Methods

1. Rinse rice with water until clean, drain well and spread evenly across a heat resistant bowl or crock pot, pour in water and curry powder.
2. Rinse squid well and gut and remove the membrane. Dice onion. Sprinkle onion over rice and place squid on top.
3. Cover the bowl with aluminum foil, or cover with a lid. Cook in oven at 200℃ for 50 minutes until done.
4. Blanch carrot, peas and corn kernels in boiling water with salt added, then remove and drain well.
5. Remove the curry rice from oven and take out the squid. Combine the rice with the three kinds of vegetables well, then season with salt and ground black pepper to taste.
6. Stuff the rice inside each squid tightly.
7. Cut the stuffed squid into sections and arrange on a serving plate. Serve.

Tips

Another easy way to prepare is to cut the squids horizontally into circles, then combine with rice and onion, then cook in oven. This way saves procedune of stuffing.

Squid Veggie Curry Rice

蘑菇起司焗飯

Mushroom & Cheese Rice

烤箱用 烤/ 模具 cookware	烘烤功能 Functions	溫度℃	時間 min
耐熱烤皿＋鋁箔紙 Heat Resistant Baking Utensils+ Aluminum Foil	or	200	50

材料

白米2杯、清水2 1/4杯、蘑菇100公克、鮮香菇100公克、洋蔥75公克、培根35公克、披薩起司100公克、帕米森起司粉35公克

❶ 無鹽奶油1小匙、鹽2小匙、黑胡椒粉1/2小匙

做法

1. 白米以份量外的清水洗淨，瀝乾水分後，放入耐熱烤皿或陶鍋內舖平，再倒入清水。
2. 洋蔥、培根切丁，蘑菇和鮮香菇切片，均勻地撒在白米上面。
3. 烤皿上面蓋上耐熱鍋蓋或鋁箔紙，放入烤箱以200℃烘烤50分鐘。
4. 取出烤好的材料，加入材料❶以飯匙翻拌均勻。
5. 將披薩起司平均擺在白飯表面，撒上起司粉，以鋁箔紙或鍋蓋蓋上，放入留有餘溫的烤箱內5分鐘後即可。

Tips 可將配方中的清水2又1/4杯，改成清水2杯、無糖動物鮮奶油1/4杯，這樣可以煮出口感香濃的米飯。材料中亦可加入少許雞肉丁，增加口感。

Ingredients

2C. white rice, 2 1/4C. water, 100g pizza cheese, 35g parmesan cheese powder, 100g cremini mushrooms, 100g fresh shiitake mushrooms, 75g onion, 35g bacon

❶ 1t. unsalted butter, 2t. salt, 1/2t. ground black pepper.

Methods

1. Rinse rice with water until clean, drain and spread evenly across a heat resistant baking bowl or ceramic pot, then pour in water.
2. Dice onion and bacon. Cut cremini mushrooms and fresh shiitake into slices, then sprinkle evenly over rice along with onion and bacon.
3. Cover the bowl with aluminum foil or with a lid. Place in oven and cook at 200℃ for 50 minutes until done.
4. Remove the bowl from oven and add ingredients ❶ to mix. Stir well with a spoon and sprinkle with pizza cheese
5. Sprinkle with cheese powders, put the aluminum foil or lid back on the bowl and return to oven. Use the remaining heat to cook until done for about 5 minutes. Remove and serve.

Tips

The 2 1/4C of water can be substituted with 2C of water and 1/4C of unsweetened whipping cream. This way the rice will be thick and fragrant. A little diced chicken can be added to enhance the texture.

Mushroom & Cheese Rice

上海青江菜飯

Shanghai Style Rice with Bok Choy

烤箱用 烤/ 模具 cookware	烘烤功能 Functions	溫度℃	時間 min
耐熱烤皿＋鋁箔紙 Heat Resistant Baking Utensils+ Aluminum Foil	or	200	40

材料

白米2杯、清水21/4杯、青江菜150公克、大蒜2瓣、橄欖油1大匙、鹽2小匙

做法

1. 將青江菜洗淨瀝乾切成末，大蒜去皮。
2. 白米以份量外的清水洗淨，瀝乾水份，放入耐熱烤皿或陶鍋內舖平，倒入清水，放入大蒜和橄欖油。。
3. 烤皿蓋上鋁箔紙或耐熱鍋蓋，放入烤箱以200℃烘烤40分鐘。
4. 取出大蒜，放入切成細末的青江菜、鹽，以飯匙攪拌均勻，將鋁箔紙或鍋蓋蓋上，放入留有餘溫的烤箱內，續烤10分鐘。
5. 取出青江菜飯即可食用。

Tips 如果要省時，可以先將青江菜和米粒混合加熱，或是先用1小匙油將青江菜炒過，再放入米粒中混合加熱亦可。

Ingredients

2C. white rice, 2 1/4C. water,150g baby bok choy, 2 cloves garlic, 1T. olive oil,2t. salt

Methods

1. Rinse baby boy choy well, drain and mince well. Remove skin from garlic.
2. Rinse rice with water until clean, drain well and spread evenly in a heat resistant baking bowl or crock pot. Pour in water and add garlic as well as olive oil.
3. Cover the bowl with aluminum foil, or cover with a lid. Cook in oven at 200℃ for 40 minutes until done.
4. Remove the bowl from oven, open the foil or the lid, discard garli and add minced bok choy. Season with salt to taste, then stir well with a spoon and put the aluminum foil or lid back to the bowl. Return to oven, use the remaining heat to cook for 10 minutes.
5. Remove the bowl and serve.

Tips

To save time, baby bok choy and rice can be cooked together. Or stir-fry baby bok choy with 1 teaspoonful of oil first, then add rice to mix.

Shanghai Style Rice
with Bok Choy

台式油飯

Taiwanese Style Oiled Rice

烤箱用 烤/ 模具 cookware	烘烤功能 Functions	溫度℃	時間 min
耐熱烤皿＋鋁箔紙 Heat Resistant Baking Utensils+ Aluminum Foil	or	200	40

材料

長糯米300公克、沙拉油3大匙、清水300公克、豬前腿肉300公克、乾香菇10小朵、栗子75公克、乾蝦米1/4杯、紅蔥頭5粒

❶ 醬油3大匙、冰糖1小匙、白胡椒粉1/2小匙、米酒1小匙

❷ 鹽1小匙、醬油1 1/2大匙、白胡椒粉1小匙

做法

1. 糯米洗淨，浸泡份量外的清水1小時，瀝乾水分，放入耐熱烤皿或陶鍋內，加入沙拉油和清水拌勻後舖平；栗子浸泡份量外的清水，再以牙籤剔除表面厚膜。
2. 豬肉切成條狀，香菇泡軟切絲，放入攪拌盆以❶料浸泡30分鐘。
3. 紅蔥頭切末，與蝦米、栗子平均撒在白米上面。
4. 豬肉和香菇瀝乾浸汁，也撒在白飯上面。
5. 烤皿上面蓋上鋁箔紙或耐熱鍋蓋，放入烤箱以200℃烘烤40分鐘。
6. 加入材料❷，用飯匙將材料拌勻，即可食用。

Tips　油飯的另外一個做法是用1大匙油將紅蔥頭、乾蝦米炒過，加材料❷拌勻，關火。接著以爆香的材料，倒入煮熟的糯米飯內拌勻亦可。

Ingredients

300g long glutinous rice, 3T. cooking oil, 300g water, 300g pork picnic shoulder, 10 small dried shiitake mushrooms, 75g chestnuts, 1/4C. dried miniature shrimp, 5 cloves shallots

❶ 3T. soy sauce, 1t. rock sugar, 1/2t. white pepper, 1t. rice wine 2 1t. salt, 1 1/2T. soy sauce, 1t. white pepper

❷ 1t. salt, 1 1/2T. soy sauce, 1t. white pepper

Methods

1. Rinse rice well with water first, then soak in water for 1 hour. Drain and spread evenly across a heat resistant baking bowl or crock pot along with cooking oil and water added. Soak chestnuts in water for some time, then remove the thick outer skin with a toothpick.
2. Cut pork into strips. Soak shiitake mushrooms in water until soft, then shred, soak these two ingredients in ingredient ❶ in a mixing bowl for 30 minutes.
3. Mince shallots and sprinkle evenly over rice along with dried shrimp and chestnuts.
4. Drain pork and mushrooms well and spread over rice.
5. Cover the baking bowl with aluminum foil, or a heat resistant lid. Cook in oven at 200℃ for 40 minutes.
6. Remove bowl, undo the foil or open the lid. Add ingredient ❷ , stir with a spoon until all the ingredients are well mixed with the rice. Serve.

Tips

Another way of preparing this dish is to stir-fry fried shallots and dried miniature shrimp with 1 tablespoon of cooking oil, add ingredient ❷ to mix, then remove from heat and add to the cooked sticky rice to mix well.

Taiwanese Style Oiled Rice

Chinese dessert & Western dessert

中西式點心

從在地台灣味到甜蜜西方風，從吃不膩的蛋糕，
到口感鮮明的甜點，烤箱菜教你快樂做糕餅！

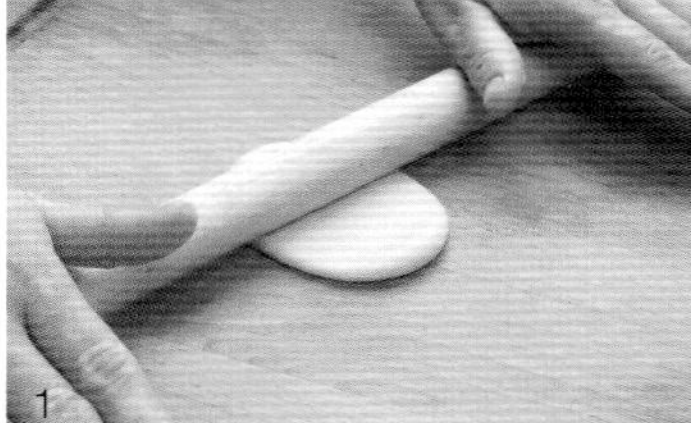

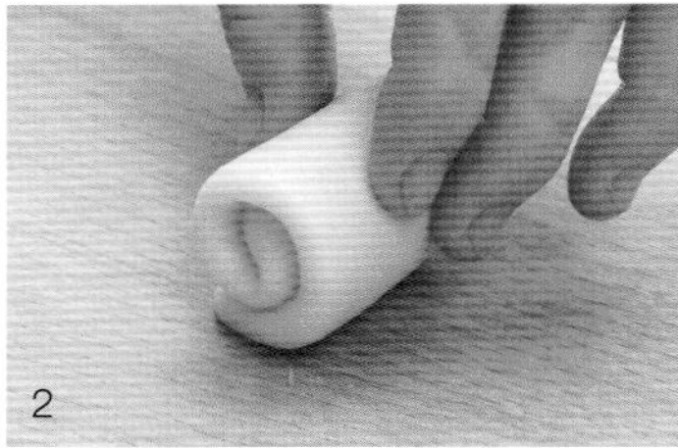

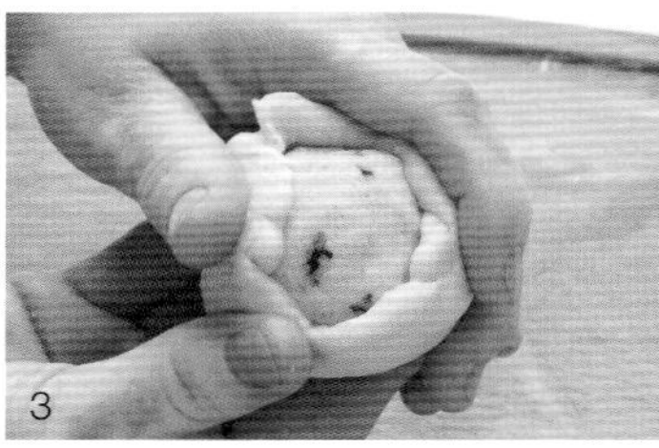

Portions 20

Ingredients

500g mung bean paste, 25g minced fried shallots, 1t. salt, 1t. sugar, 90g fatty pork, 15g roasted white sesame seeds, 1T. curry powder

❶ Sweet pastry dough: 250g all purpose flour, 10g confectioner's sugar, 100g anhydrous milk fat (butter oil/shortening), 75g lukewarm water (approximately 35C), 1/2t. salt

❷ Crispy pastry dough: 200g cake flour, 100g anhydrous milk fat(butter oil/shortening)

咖哩綠豆凸

Bean Stuffed Curry Pastry

烤箱用 烤/ 模具 cookware	烘烤功能 Functions	溫度°C	時間 min
耐熱烤盤 Heat Resistant Baking Sheet		180	25

份量 20個

材料

綠豆沙500公克、紅蔥頭末25公克、鹽1小匙、糖1小匙、肥豬肉90公克、熟白芝麻粒15公克、咖哩粉1大匙

❶ 油皮：中筋麵粉250公克、糖粉10公克、脫水奶油(酥油)100公克、溫水(約35℃)75公克、鹽1/2小匙

❷ 油酥：低筋麵粉200公克、脫水奶油(酥油)100公克

做法

1. 將肥豬肉與紅蔥頭末放入炒鍋中炒熟，加入咖哩粉拌勻，起鍋與所有材料混合拌勻成內餡，再分成每個40公克，搓圓後備用。
2. 將油皮材料❶的麵粉和糖粉混合過篩後，放置於工作檯築成粉牆，也就是在粉料的中間撥出一個洞，放入油、鹽以及溫水，混合後搓揉成光滑的麵糰，蓋上保鮮膜鬆弛30分鐘，再分成每個20公克的小麵糰，搓圓備用。
3. 將油酥材料❷的低筋麵粉過篩後混合脫水奶油，以刮麵刀用切、拌、按、壓的方式拌成麵糰，再分成每個15公克的油酥，搓圓備用。
4. 取一個油皮包入一個油酥，收口捏緊並朝上，以擀麵棍擀成長形的薄片，由下往上捲起，接合處朝上，再擀平後捲起，蓋上保鮮膜鬆弛20分鐘。(圖1、圖2)
5. 取出麵糰擀成中間厚周圍薄的圓形片，包入咖哩綠豆餡，收口捏緊朝下放置在烤盤上。此時開啓烤箱預熱，放入麵糰以180℃烘烤25分鐘。(圖3)

Methods

1. Stir fry pork and fried shallot in pan until done, add curry powder to mix. Remove from heat and combine with all the ingredient to make fillings, then divide the fillings into portions, 40 grams each, and roll each portion into a round ball.
2. Combine flour and confectioner's sugar from ingredient ❶ together, then sift and place on a working surface. Dig a hole in the center and build a flour surround. Add shortening, salt and lukewarm water in center, slowly and gently fold and combine together. Roll and knead into a soft smooth dough. Cover with saran wrap and relax for 30 minutes and divide into small portions, 20 grams each, then roll each portion into a round sweet pastry ball.
3. Sift cake flour from ingredient ❷ and mix well with shortening. Use a dough scraper to cut, mix, press and squeeze until the dough is incourpreated, then divide the dough into portions, 15 grams each. Roll each portion into a round pastry ball.
4. Hold the pastry dough in the hand and press flat, then place the sweet pastry ball in center. Wrap up and seal tightly with the opening facing up. Roll with a rolling pin into a rectangle thin skin, then roll from bottom up into a cylinder with the opening facing up. Roll again and roll from bottom up once again, then cover with a saran wrap to relax for 20 minutes. Repeat this process to finish all the dough. (fig 1.2)
5. Place a piece of dough on the working surface, roll the dough into a round flat skin with sides thinner than the center. Wrap the curry filling in the center, seal tight and arrange on the baking sheet with the opening facing down. Repeat the process until all the pastry is done. Preheat oven. Remove the curry pastry to oven and bake at 180℃ for 25 minutes until done. (fig 3)

1

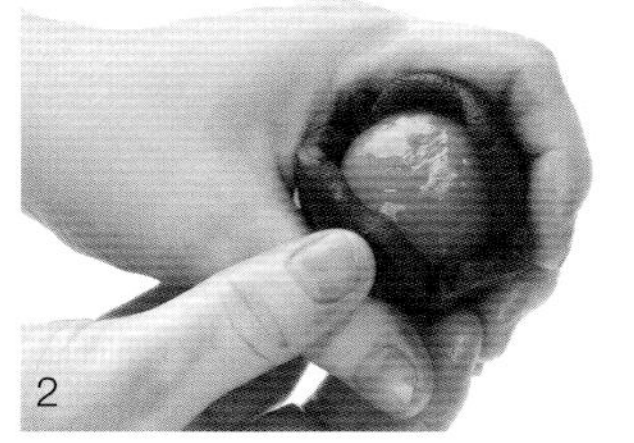
2

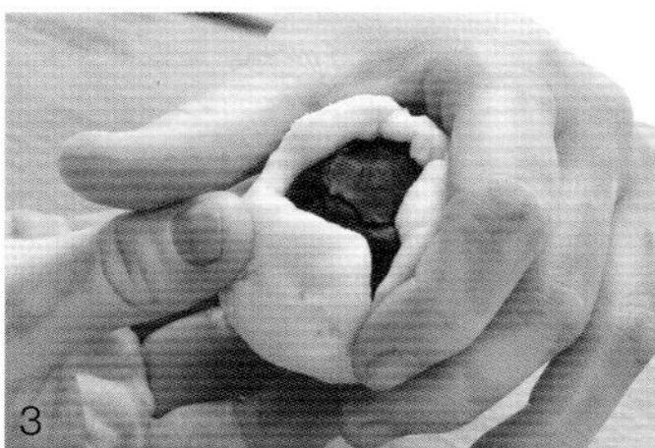
3

Portions 20

Ingredients

20 duck egg yolks, 1 bowl Kaoliang wine, 100g black sesame seeds, 600g red bean paste, egg yolk liquid as needed for brushing over pastries

❶ Sweet Pastry Dough: 200g all purpose flour, 40g confectioner's sugar, 70g anhydrous milkfat (butter oil/shortening), 1/2t. salt, 65g lukewarm water (approximately 35℃)

❷ Crispy Pastry Dough: 200g

蛋黃酥

Bean Paste with Egg Yolk Pastry

烤箱用 烤/ 模具 cookware	烘烤功能 Functions	溫度℃	時間 min
耐熱烤盤 Heat Resistant Baking Sheet		180	25

份量：20個

材料：

鴨蛋黃20顆、高粱酒1碗、黑芝麻粒100公克、烏豆沙餡600公克、塗抹用蛋黃液適量

❶ 油皮：**中筋麵粉200公克、糖粉40公克、脫水奶油(酥油)70公克、鹽1/2小匙、溫水(約35℃)65公克**

❷ 油酥：**低筋麵粉200公克、脫水奶油(酥油)100公克**

做法：

1. 將鴨蛋黃沾裹高粱酒，或是將高粱酒裝入灑水瓶，以噴灑方式將高粱酒噴灑在蛋黃上。把蛋黃排入烤盤，放入烤箱以200℃烘烤10分鐘至熟。取出後，以電風扇吹蛋黃幫助降溫。
2. 將烏豆沙餡分成每個30公克，每個餡包入1顆鴨蛋黃。
3. 做法請參照p.88的做法2.3.4
4. 取出麵糰擀成中間厚、周圍薄的圓形片，包入豆沙蛋黃餡，收口捏緊朝下放置在烤盤上。麵糰表面塗抹蛋黃液，撒上少許的黑芝麻。此時開啓烤箱預熱，放入麵糰以180℃烘烤25分鐘。

Methods

1. Coat duck egg yolks with wine, or fill sprinkler with wine and sprinkle over duck egg yolks.
2. Arrange duck egg yolks in rolls on a baking sheet and cook at 200℃ in oven for 10 minutes until done. Remove and blow with a fan to lower the temperature.
3. Divide the bean paste into equal portions of 30g each, press flat and place duck egg yolks in center of each bean paste portion, then wrap up well.
4. Sift flour and confectioner's sugar from ingredient ❶ and place on a working surface. Make a hole in the center and build up a wall of flour. Add shortening, salt and lukewarm water in the center, then gently mix together with the flour. Knead and roll into a smooth, soft sweet pastry dough and cover with saran wrap. Let rise for 30 minutes, then divide into small portions of 20g each and roll each portion into a round ball with the palms of the hands.
5. Sift the flour from ingredient ❷ and mix well with the shortening. Use a scraper to cut, mix, press, and squeeze until the crispy pastry dough is smooth and elastic, then divide the dough into equal portions of 15g each. Roll each portion into a round ball with the palms of the hands.
6. Wrap each crispy pastry dough ball up inside the sweet pastry dough, squeeze the opening tight and place face up, then roll into a thin rectangular shaped wrapper with a rolling pin. Roll the wrapper from the bottom up into a cylinder with the opening face up, then roll flat and roll up into a cylinder again. Cover with saran wrap and let rest for 20 minutes.
7. Use a rolling pin and roll the cylinder into round skin with sides thinner than the center. Place the stuffed bean paste in the center and wrap up well. Seal the opening tight and place upside down on a baking sheet. Brush the surface of the pastry with egg yolk liquid and sprinkle with little black sesame seeds on top. Repeat to finish all the pastries. Preheat oven and bake the stuffed pastries at 180C for 25 minutes until done.

香草磅蛋糕

Vanilla Pound Cake

烤箱用 烤/ 模具 cookware	烘烤功能 Functions	溫度℃	時間 min
耐熱烤模 Heat Resistant Baking molds		175	45

材料

無鹽奶油225公克、糖粉225公克、雞蛋200公克、低筋麵粉250公克、泡打粉1小匙、香草條1/2支

做法

1. 香草條縱向切半，以小刀刮出裡面的黑色香草籽，將香草籽抹在奶油上面。
2. 奶油切小塊放入攪拌盆中，置室溫下軟化。
3. 糖粉過篩後加入攪拌盆中，以攪拌器將奶油和糖粉打勻。
4. 雞蛋分5至6次慢慢加入奶油糊中攪拌。
5. 最後將粉類過篩加入攪拌盆，改以橡皮刮刀拌勻，即成麵糊。
6. 此時開啟烤箱預熱，並準備烤模。在長方形烤模內側舖一張烘焙紙，或是塗抹份量外的奶油並撒上高筋麵粉亦可。
7. 將麵糊倒入烤模中，形成兩側略高、中央凹陷的狀態，放入烤箱以175℃烘烤45分鐘即可。

Tips 雞蛋打散後，要分次且緩慢地加入奶油糊中攪拌。因為奶油和雞蛋不易融合，如果一次加入大量的雞蛋，恐怕無法將材料混合均勻，同時又產生油水分離的現象。

Ingredients

225g unsalted butter, 225g confectioner's sugar, 200g eggs, 250g cake flour, 1t baking powder, 1/2 vanilla bean pod

Methods

1. Halve vanilla bean pod horizontally, then scrape out the black vanilla seeds inside with a small knife. Spread the seeds on butter.
2. Cut butter into small pieces and remove to a mixing bowl, then let stand at room temperature until soft.
3. Sift confectioner's sugar and add the mixing bowl, use a mixer and beat butter and sugar until evenly mixed.
4. Add eggs little at a time for about 5-6 times to the butter mixture.
5. Sift flour and fold into the mixing bowl, use a rubber spatula and mix well to form flour batter.
6. Preheat oven and prepare a baking mold. Line a sheet of parchment paper at the bottom of the rectangle mold. Or grease the mold with butter and sprinkle with bread flour.
7. Pour the flour batter into the mold to form the sides are higher than the center. Place in oven and bake at 175℃ for 25 minutes until done.

Tips

Break eggs into bowl first, then divide into several portions to add to the butter because butter and eggs do not combine easily. If eggs are added all at one time, it will be difficult to combine the ingredients well together, and cause the oil to separate from the liquid.

Vanilla Pound Cake

核桃布朗尼

Walnut in Brownies

烤箱用 烤/模具 cookware	烘烤功能 Functions	溫度℃	時間 min
耐熱圓形烤模 Heat Resistant Round Baking Molds		150	40

材料

苦甜巧克力340公克、無鹽奶油115公克、雞蛋3顆、蘭姆酒1大匙、核桃碎75公克

❶ 低筋麵粉115公克、泡打粉1/2小匙、鹽1/4小匙

做法

1. 苦甜巧克力切碎放入鋼盆，奶油也放入鋼盆，兩種材料以隔水加熱的方式融化。
2. 融化後離火，分次打入雞蛋，用電動攪拌器拌勻。
3. 將材料❶混合過篩加入，以橡皮刮刀拌勻，最後倒入藍姆酒拌勻即成麵糊。
4. 此時開啟烤箱預熱，並準備烤模，模型底部舖一張烘焙紙，周圍塗抹份量外的奶油。
5. 將麵糊倒入模型中整平，核桃碎排列在表面並輕輕壓入，將模型放入烤箱，以150℃烘烤40分鐘。

Tips 建議淋上打發過的無糖動物鮮奶油品嘗，味道會更好。

Ingredients

340g bittersweet chocolate, 115g unsalted butter, 3 eggs 1T. rum, 75g chopped walnuts

❶ 115g cake flour, 1/2t. baking powder, 1/4t. salt

Methods

1. Chop bittersweet chocolate into pieces and remove to a stainless pan along with butter added. Melt these two ingredients with a double boiler until completely melted.
2. Remove from heat, add eggs, a little at a time and beat with an electric mixer until even.
3. Combine ingredient ❶ and sift first before adding to method (2), scrape with a rubber spatula until well-mixed, then pour in rum to mix. The batter is done.
4. Preheat oven and prepare a baking mold. Line the bottom of the mold with a sheet of parchment paper, then grease the sides with extra butter.
5. Pour the batter inside the mold and spread it evenly. Place the walnuts on top by pressing lightly into the batter. Place in oven and bake 150℃ for 40 minutes until done.

Tips

Drizzle with unsweetened whipped cream on top if desired, which will enhance the flavor even more.

Walnut in Brownies

美式巧克力餅

American Style Chocolate Cookie

烤箱用 烤/模具 cookware	烘烤功能 Functions	溫度°C	時間 min
耐熱烤盤 Heat Resistant Baking Sheet		200	12～15

材料

無鹽奶油150公克、雞蛋100公克、裝飾用糖粉100公克

❶ 低筋麵粉275公克、糖粉150公克、小蘇打粉1/2小匙、可可粉25公克、肉桂粉1/2小匙

做法

1. 將奶油放入鋼盆，置室溫下軟化。烤盤舖烘焙紙。此時開啓烤箱預熱。
2. 將材料❶混合過篩放入鋼盆，加入雞蛋以攪拌器慢速攪拌均勻成麵糰。
3. 用冰淇淋勺將麵糰挖成20等份，排在烤盤上，撒上糖粉，放入烤箱以200℃烘烤12至15分鐘。

Tips 配方中還可以加入75公克的碎核桃粒或是杏仁粒，增加口感的豐富度。

Ingredients

150g unsalted butter, 100g eggs, 100g confectioner's sugar for garnishing

❶ 275g cake flour, 150g confectioner's sugar, 1/2t. baking soda, 25g cocoa, 1/2t. cinnamon powder

Methods

1. Place butter in a stainless bowl and let stand at room temperature until soft. Line baking sheet with a sheet of parchment paper. Preheat oven.
2. Combine ingredient ❶ well together, then sift and remove to the bowl, along with eggs added. Use a mixer and beat slowly until evenly mixed to form a cookie dough.
3. Use an ice cream scoop to scoop out 20 equal portions of dough, arrange on the baking sheet and sprinkle with confectioner's sugar. Place in oven and bake at 200℃ for 12 to 15 minutes until done.

Tips

75 grams of chopped walnuts or whole almonds can be added to the recipe to increase the richness of the texture.

American Style Chocolate Cookie

紅茶香蕉瑪芬

Banana & Tea Muffins

烤箱用 烤/ 模具 cookware	烘烤功能 Functions	溫度°C	時間 min
5.5cm耐熱烤模x1 5.5cm Heat Resistant Baking Moldsx1		175	25

份量　約18個

材料

無鹽奶油110公克、細砂糖110公克、全蛋2顆、低筋麵粉250公克、泡打粉1小匙

❶ 牛奶35公克、香蕉泥75公克、紅茶粉末1大匙

做法

1. 奶油和糖放入攪拌盆，用攪拌器打至非常均匀。
2. 將蛋打散，分3至4次加入攪拌盆混合均匀。
3. 將低筋麵粉和泡打粉混合過篩，加入攪拌盆，此時改用橡皮刮刀將材料混合。
4. 最後加入材料❶拌匀即成麵糊。此時開啓烤箱預熱，準備瑪芬烤模。
5. 將麵糊填入烤模內，並將每個模型輕敲桌面數下，讓麵糊可以填滿整個烤模。
6. 將模型放入烤箱，以175℃烘烤25分鐘。

Tips　可以改用蘋果或水蜜桃來取代香蕉，或是只單純使用紅茶來調味亦可。

Portions

approximately 12

Ingredients

110g unsalted butter, 110g fine granulated sugar, 2 whole eggs, 250g cake flour, 1t. baking powder

❶ 35g milk, 75g mashed banana, 1T. ground black tea

Methods

1. Beat butter and sugar in a mixing bowl and until evenly mixed.
2. Break eggs and add to the mixing bow , a little at a time, approximately 3-4 times until evenly mixed.
3. Combine cake flour and baking powder well together, then fold into the mixture. Use a spatula and mix well.
4. Add ingredient ❶ to form batter. Preheat oven and prepare muffin pans.
5. Fill the muffin pans with batter and shake each pan against the working surface couple of times to let the batter fill the pans completely.
6. Remove the pans to oven and bake at 175℃ for 25 minutes until done.

Tips

Banana can be substituted with apple or peach, or just use pure black tea as main ingredient.

Banana & Tea Muffins

香濃起司蛋糕 Cheese Cake

烤箱用 烤/模具 cookware	烘烤功能 Functions	溫度°C	時間 min
耐熱烤模 Heat Resistant Baking Molds		160	50

材料

奶油乳酪400公克、酸奶油100公克、細砂糖100公克、雞蛋4顆、柳橙汁50公克、柳橙皮1大匙、融化奶油35公克

做法

1. 奶油乳酪切小塊放入攪拌盆，置室溫下軟化。
2. 將細砂糖加入攪拌盆，以攪拌器將材料攪拌至均勻混合。
3. 加入酸奶油拌勻。
4. 雞蛋分四次加入拌勻，攪拌的過程必須停下機器，將附著在盆邊的材料刮下，以免出現材料攪拌不均的狀況。
5. 最後加入柳橙汁、柳橙皮和融化奶油拌勻，即成麵糊。
6. 將麵糊透過細目型的篩網過濾，使麵糊的組織更細緻。
7. 此時開啓烤箱預熱，並準備烤模。烤模底部和周圍以鋁箔紙包住，模型內塗抹份量外的奶油。
8. 準備一個有深度的烤盤，倒入熱水並放置烤模；將麵糊倒入烤模，放入烤箱以160℃烘烤50分鐘。
9. 取出烤好的起司蛋糕，待降溫後，放入冰箱冷藏，至完全冰透後，將烤盤周圍以熱毛巾包裹，再將蛋糕取出即可。

Tips　柳橙皮可以切成細末後再加入，或是只加入柳橙汁而不加柳橙皮亦可。

Ingredients

400g cream cheese, 100g sour cream, 100g fine granulated sugar, 4 eggs, 50g orange juice, 1T. orange peel, 35g melted butter

Methods

1. Cut cream cheese into small pieces and remove to a mixing bowl. Let sit at room temperature until softened.
2. Add granulated sugar and beat until the ingredients are evenly mixed.
3. Add sour cream to mix.
4. Divide eggs into 4 parts and add to the mixture slowly. During the beating process, stop once a while to scrape the mixture from the sides of bowl to prevent any uneven mixture.
5. Add orange juice, orange peel and melted butter. The batter is formed.
6. Pour the batter through a sieve to make the texture even finer.
7. Preheat oven and prepare a baking mold. Line the bottom and sides of the mold with a sheet of aluminum foil, then grease with extra butter.
8. Prepare a deep baking pan and fill with steaming water. Place the baking mold inside the pan. Fill the mold with batter and bake in oven at 160℃ for 50 minutes until done.
9. Remove the cake and wait until the temperature drops before placing in the refrigerator. Chill until icy cold, surround the sides with a thick steaming towel, then flip it upside down to remove the cake from mold. Serve.

Tips

Orange peel can be minced finely before being added into the mixture, or just add the orange juice without the peel.

Cheese Cake

Tips

製作派皮時，使用冰過的奶油塊，可以讓烤好的派更香酥。

Tips

Use frozen butter to make the pie shell, so that the baked pie tastes even more crispy and delicious.

杏仁蘋果派 Almond Apple Pie

烤箱用 烤/模具 cookware	烘烤功能 Functions	溫度℃	時間 min
耐熱烤盤 Heat Resistant Baking Sheet	[icons]	180、175	15、150

材料

杏仁片3大匙、裝飾用糖粉適量、蘋果餡

❶ 派皮：中筋麵粉300公克、杏仁粉30公克、冰水60～70c.c.、糖粉40公克、鹽1小匙、無鹽奶油200公克、全蛋1顆

做法

1. 將材料❶的粉類倒在工作檯上，築成一個粉牆。粉牆內打入雞蛋，並混合部份的粉成爲小麵塊，加入冰凍的奶油，以切麵刮刀用切、拌、按、壓的方式拌成麵糰。麵糰覆上保鮮膜，置室溫下鬆弛20分鐘。
2. 準備派模，模型裡面塗上奶油。將麵糰放置在工作檯上，用擀麵棍擀成比派模還大的薄片麵皮，將麵皮放入派模中，輕輕壓入貼緊，多餘的派皮以切麵刮刀切除。（圖1）
3. 取一支叉子，在派皮的表面搓滿洞，蓋上一張鋁箔紙，紙上放滿烘焙用重石或是豆子，將派皮放入烤箱，以180℃烘烤15分鐘。取出派，將煮好的蘋果餡料放入填滿。（圖2）、（圖3）
4. 將多餘的派皮揉成糰，以擀麵棍擀平，再切成寬度約1.5公分的長條，交叉排列在派的表面，多餘的部份以切麵刮刀切除。
5. 準備蛋液，用毛刷將蛋液塗在派皮表面，並黏上杏仁片。此時開啓烤箱預熱。
6. 把蘋果派放在烤盤上入烤箱，以175℃烘烤15分鐘。取出蘋果派略待降溫後，表面撒上糖粉即可切片食用。

Ingredients

3T. almond flake, confectioner's sugar as desired for garnishing,Apple filling(1200g fresh peeled, seeded apples, 100g fine granulated sugar, 30g lemon juice, 30g unsalted butter)

❶ Pie shell: 300g all purpose flour, 30g almond powder, 60～70c.c , 40g confectioner's sugar, 1t. salt, 200g unsalted butter, 1 whole egg

Methods

1. Place flour from ingredient ❶ on a working surface, dig a hole in center and build a flour wall around. Break the egg inside and slowly combine the flour with the eggs, then add frozen butter. Use a dough scraper to cut, mix, press and squeeze the dough until it is in coorperated.Cover the dough with a saran wrap and rest at room temperature for 20 minutes.
2. Prepare a pie pan and grease the pan with butter. Place the dough on the working surface, roll out the dough with a rolling pin into a flat, large pie skin, bigger than the pie pan. Place the pie skin onto the pie pan, press down into the pan gently and tightly, then trim off any excess pie skin with the dough scraper.(fig.1)
3. Get a fork and poke holes on surface, cover with an aluminum foil sheet and arrange baking"rocks" or "beans", then bake at 180℃ for 15 minutes.Remove pie from oven and discard the aluminum baking rocks/beans, then fill the center with cooked apple filling. (fig.2,3)
4. Roll out the remain dough flat, then cut into long strips about 1.5cm wide, then line in to Trim off any excess strips.
5. Brush the beaten eggs on the pie surface, then top with almond flakes. Preheat oven.
6. Place apple pie on the baking sheet and bake in oven. Bake at 175℃ for 15 minutes until done, then remove and wait until the pie cools down. Dust confectioner's sugar and cut into slices and serve.

1

2

3

Egg Tarts

Tips
可以改用鮮奶來代替奶水，或是使用無糖動物鮮奶油。

Tips
Evaporated milk can be substituted with fresh milk, or unsweetened whipping cream.

蛋塔 Egg Tarts

烤箱用 烤/ 模具 cookware	烘烤功能 Functions	溫度℃	時間 min
6.5 cm 耐熱烤模 x12 6.5cm Heat Resistant Baking Molds x12		150	50

份量 約12個

材料

全蛋1顆、鹽1/4小匙、無鹽奶油100公克

❶ 塔皮：低筋麵粉150公克、奶粉50公克、糖粉50公克

❷ 內餡：水100公克、細砂糖55公克、全蛋2顆、蛋黃1個、奶水100公克、香草精1小匙

做法

1. 將材料❶過篩在工作檯上，築成一個粉牆，倒入全蛋和鹽混合。
2. 混合至材料變成小顆粒狀時，加入冰凍的奶油，以切麵刮刀用切、拌、按、壓的方式拌成麵糰。
3. 麵糰覆上保鮮膜，放置室溫下鬆弛20分鐘。
4. 此時準備內餡，將水、糖倒入小鍋，加熱至糖融化即可關火。
5. 將全蛋、蛋黃、奶水和香草精混合放入攪拌盆拌勻，再倒入糖水拌勻。
6. 用細目型的篩網過濾內餡，直到表面沒有泡沫爲止。(圖1)
7. 取出麵糰分割成每個30公克的小麵糰，將麵糰壓入塔模中，塔皮盡量厚薄一致，多出的塔皮以切麵刮刀切除。(圖2)
8. 此時開啓烤箱預熱，將整型完成的塔皮放在烤盤上，塔皮內倒入內餡，入烤箱以150℃烘烤50分鐘。(圖3)

Tips　可以改用鮮奶來代替奶水，或是使用無糖動物鮮奶油。

Portions

approximately 12

Ingredients

1 whole egg, 1/4t. salt, 100g unsalted butter

❶ Tart Shell: 150g cake flour, 50g milk powder, 50g confectioner's sugar

❷ Filling: 100g water, 55g fine granulated sugar, 2 whole eggs, 1 egg yolk, 100g evaporated milk, 1t vanilla extract

Methods

1. Sift ingredient ❶ on the working surface, dig a hole in center and build up a flour wall surrounding the hole. Pour in whole egg and salt to taste.
2. Combine the flour and egg slowly to form small lumps, then add frozen butter. Use a dough scraper to cut, mix, press and squeeze until the dough is incoorporated.
3. Cover the dough with saran wrap and relax at room temperature for 20 minutes.
4. Meanwhile prepare the filling, combine water and sugar in a small pan. Cook over low heat until sugar dissolves, then remove from heat.
5. Combine whole eggs, egg yolk, evaporated milk and vanilla extract in a mixing bowl, then add sugar water to mix.
6. Pour the filling through a fine sieve until there are no bubbles on the surface.(fig.1)
7. Remove the dough and divide into small portions of dough weigh 30 grams each. Press the dough inside the tart mold, try to make the shell even, then trim off the excess tart skin with a dough scraper. (fig.2)
8. Preheat oven. Place the tarts on a baking sheet and fill each tart with fillings, then bake in oven at 150℃ for 50 minutes until done. Remove and serve.(fig.3)

Tip

Evaporated milk can be substituted with fresh milk, or unsweetened whipping cream.

雞蛋布丁

Egg Pudding

烤箱用 烤/ 模具 cookware	烘烤功能 Functions	溫度℃	時間 min
4 cm 耐熱烤模 x4 4cm Heat Resistant Baking Molds x4		175	30

份量 約4個

材料

細砂糖100公克、水1大匙

❶ 無糖動物鮮奶油150公克、牛奶150公克、細砂糖60公克

❷ 香草精1/2小匙、雞蛋150公克

做法

1. 先製作焦糖：將細砂糖放入鍋中，倒入水浸濕細砂糖，開中火煮至焦褐色後關火，立刻將糖倒入布丁模型杯底，此時杯子會變得很燙，務必留意以免燙傷。
2. 將材料❶放入小鍋邊加熱邊攪拌，煮至糖融化即可關火。
3. 把材料❷加入拌勻，即成布丁蛋液。
4. 布丁蛋液用細目篩網過濾，以去除表面的泡沫。
5. 此時開啓烤箱預熱，將布丁蛋液倒入杯中，放入烤箱以隔水加熱的方式以175℃烘烤30分鐘。
6. 把烤好的布丁取出降溫，放入冰箱冷藏。食用前以小刀沿著模型劃一圈，倒扣脫模。

Tips　隔水烘烤的布丁口感非常軟嫩，是大人小孩都喜歡的口感。

Portions

approximately 4

Ingredients

100g fine granulated sugar, 1T. water

❶ 150g unsweetened butter, 150g milk, 60g fine granulated sugar

❷ 1/2t. vanilla extract, 150g eggs

Methods

1. To prepare caramel sugar: Soak fine granulated sugar in pan first, then cook over medium heat until brown and remove from heat. Pour immediately into the bottom of the pudding molds. The molds will be extremely hot, be careful not to burn yourself.
2. Cook ingredient ❶ in pan over low and stir at the same time until the sugar dissolves, then remove from heat.
3. Add ingredient ❷ and mix well. This is pudding liquid.
4. Pour the liquid through a sieve to remove any lumps and bubbles on surface.
5. Preheat oven. Pour the liquid into the molds and place on baking sheet full of water. Place in oven and bake at 175℃ for 30 minutes until done.
6. Remove puddings and wait until puddings cool, then refrigerate until cold. Before serving, use a small knife and cut along side the rim once, flip over and remove the pudding from the mold. Serve.

Tips

The texture of the double-boiled puddings is very tender and soft, and is a favorite of both adults and children.

Egg Pudding

SHIH LEI
世磊實業
台 北 Tel:02 2760 6096
台 中 Tel:04 2452 4948
高 雄 Tel:07 3755 262
e-mail:shihlei@ms62.hinet.net
http: //www.shih-lei.com.tw
請洽全國各大廚具公司
設計攝影 光域 2767 4552

烤箱新手的第一本書

飯、麵、菜、湯品與甜點統統搞定

作者　王安琪
攝影　徐博宇
美術設計　鄭雅惠
文字編輯　彭思園
企畫統籌　李橘
總編輯　莫少閒
出版者　朱雀文化事業有限公司
地址　台北市基隆路二段 13-1 號 3 樓
電話　（02）2345-3868
傳真　（02）2345-3828
劃撥帳號　19234566 朱雀文化事業有限公司
e-mail　redbook@ms26.hinet.net
網址　http://redbook.com.tw
總經銷　大和書報圖書股份有限公司（02）8990-2588
ISBN　978-986-6780-16-5
初版三十二刷　2018.10
定價　280 元
出版登記　北市業字第 1403 號

國家圖書館出版品預行編目

烤箱新手的第一本書：
飯、麵、菜、湯品與甜點統統搞定
王安琪 著 .一初版一台北市：
朱雀文化，2007〔民 96〕
面； 公分，--（Cook50；084）
ISBN 978-986- 6780-16-5（平裝）
1. 食譜
427.1　　96024306

About 買書：

●朱雀文化圖書在北中南各書店及誠品、金石堂、何嘉仁等連鎖書店均有販售，如欲購買本公司圖書，建議你直接詢問書店店員。如果書店已售完，請洽本公司。

●●至朱雀文化網站購書（http://redbook.com.tw），可享 85 折優惠。

●●●至郵局劃撥（戶名：朱雀文化事業有限公司，帳號 19234566），掛號寄書不加郵資，4 本以下無折扣，5 ～ 9 本 95 折，10 本以上 9 折優惠。